TICKETS FOR THE ARK

FROM WASPS TO WHALES – HOW DO WE CHOOSE WHAT TO SAVE?

First published in Great Britain in 2022 by

Profile Books
29 Cloth Fair, Barbican, London EC1A 7JQ.

www.profilebooks.com

1 3 5 7 9 10 8 6 4 2

Typeset in Adobe Garamond Pro and LL Rubber Grotesque
to a design by Henry Iles.

A CIP catalogue record for this book is available from the
British Library.

ISBN 978-1788167079
eISBN 978-1782838067

The author would like to acknowledge assistance from
the British Ecological Society

Printed and bound in Great Britain by Clays Ltd, Elcograf S.p.A.
on Forest Stewardship Council (mixed sources) certified paper.

TICKETS FOR THE ARK

FROM WASPS TO WHALES – HOW DO WE CHOOSE WHAT TO SAVE?

REBECCA NESBIT

P

PROFILE BOOKS

CONTENTS

INTRODUCTION

WHY PROTECT NATURE?

Headlines about nature spark strong emotions: coral reefs are destroyed, mountain gorillas bounce back, ringed parakeets are culled. We react with despair and hope to these situations because we want *more nature*. But rarely do we take a step back and ask why. Why do changes in nature matter? What's the point of conservation? Does it justify the prices paid by people and animals? Is it ever acceptable to kill wild animals because they live somewhere we find inconvenient?

Even if we feel confident in our answers, a closer look may reveal that our reasoning doesn't stand up to scrutiny. After all, we can't be guided by science alone. Science is vital for ensuring we achieve our objectives, but it is ethical values that need to underpin them. Take honeybees, for example: science can tell us the impact of honeybee declines and what can be done about them, but it can't tell us whether honeybees are the best focus for conservation funds. They are important as pollinators for crops, but their conservation will do nothing to prevent extinctions. Which is more important to us? As we will see, the accepted narrative is not so straightforward.

Questions about what we value in nature shouldn't be solely the domain of experts, either. We are used to having a voice in ethical debates around medical science, such as stem cell research or 'three-parent IVF'. Personal values are as relevant to conservation as they are to medicine – we all have a role to play in making decisions. Everyone has values that are relevant for conservation and visions of the world we want to create. The arguments between supporters of salmon and seals in Scotland's Moray Firth, in Chapter 8, reveal how vital it is to hear different voices – in that case from the fishing industry, conservationists and tour operators.

All too often we see conservation answers as so startlingly obvious that we don't question them. But we can't simply return to a past state of nature in this age of unprecedented extinction. As the story of the Guam rail louse in Chapter 4 shows, some species are gone for good – and when we lose a species we lose, too, its impact on the ecosystem. So, as the natural world changes, we need to consider fundamental questions. Are the species we love the best ones to protect? Do all extinctions matter? When should we welcome species in new ranges? Is it ever right to kill wild animals? What sacrifices should we make in the name of conservation?

These are questions we need to answer if conservation is to bring the greatest benefits. The answers will allow us to allocate funds wisely and make laws which are fair and effective. Even if we don't agree on the answers, an understanding of why we hold our beliefs is invaluable when resolving conservation debates. In Chapter 5, we look at a decision obvious to most conservationists – culling rats to prevent the extinction of the Floreana mockingbird – but one opposed by animal rights advocates. Clear answers allow everyone to negotiate a way forwards.

The stakes are high – our current attitude to nature can, literally, lead us towards disaster. This needn't be the outcome: humans can still thrive in a changing world, as can other species. A first step is to address the challenging questions surrounding what we want from conservation – questions that are both uniquely modern and rooted in ancient mythology. The flood story of Noah's Ark has captured imaginations for millennia. There are similarities between the story of the Mesopotamian flood and what is happening now – the Earth is changing and, if we don't act, then species will go extinct, maybe even humans. But in the ancient myths, the animals were kind enough to present themselves so they could board the Ark, yet today there are species going extinct without humans even knowing about them.

The resources we dedicate to conservation will never be enough to prevent all extinctions, and we are forced to choose our priorities. We may commit resources to saving one species in the knowledge that others are condemned to extinction. And allocation of limited funds isn't the only way that conservation picks winners and losers. When one species thrives, it can lead to the decline in another. Noah's animals were well behaved – they refrained from eating each other or allowing their offspring to take up too much space.

In reality, every single organism must eat and procreate at another's expense. Conservation is therefore always about making trade-offs, supporting one species at the cost of others. These dilemmas will form the basis of this book. In each chapter we will pit one species against another to tackle the essential questions that conservation must face.

CHAPTER 1

THE MYTH OF WILD NATURE

Bison vs Siberian larch (and human interventions from reefs to pigeons)

It took Nikita Zimov thirty-five days to transport a dozen bison from Denmark to the Siberian Arctic – a 13,000-kilometre journey by truck, fording rivers and crossing mountains on stony tracks, then 1,500 kilometres on a barge. It was the realisation of a dream that his father had nurtured for twenty years and is part of their plan to fight global warming by destroying trees – a bold vision which has captured the attention of scientists and the public. It has led them through some creative ideas, from crowdsourcing the money to transport the bison (successful) to buying a military tank for clearing trees (less successful).

The Zimovs' ultimate ambition is to increase the amount of carbon stored in the Arctic, and they are starting with their patch of land in eastern Russia. Their plan involves some serious ecological engineering and it is inspired by the Arctic landscape as it was before the arrival of humans. Trees such as the Siberian larch were once kept in check by herbivores such as rhinos, steppe bison, elk and the woolly mammoth, but as the last ice age drew to a close these magnificent beasts disappeared from the fossil record. We still haven't agreed what caused their extinctions. Was it climate? Was it people? Was it a combination of both? The debate will no doubt continue, but the fact remains that the world's ecosystems have been dramatically altered by their loss.

Mammoth steppe once covered vast areas of land in Europe, Asia and North America, and this habitat disappeared along with the herbivores. Where Siberia was once dominated by grasses and herbs, it is now covered in tall shrubs, sparse trees and a carpet of moss. The herbivores had maintained the steppe by cycling nutrients and trampling vegetation. Without them, grasses and herbs couldn't compete with moss or trees. Siberian larch triumphed.

Sergey Zimov, the visionary Russian scientist who dreamt up the idea of restoring a prehistoric landscape, explains his mission: 'During the ice age, there were millions and millions of herbivores. Now we want to bring them back. We are not trying to establish a new ecosystem – we're only trying to reconstruct the ecosystem which existed here 13,000 years ago.'

When Sergey began his quest, he had many doubters, but thankfully he managed to convert the most important person: his son Nikita. When he turned 20, Nikita had been planning to leave for the city, but he is instead dedicating his life to his father's vision of restoring the mammoth steppe. Sergey and

Nikita have dubbed their small patch of land Pleistocene Park, to reflect the geological era they hope to recreate. It lies on the bank of the Kolyma River, a 45-kilometre boat journey south of the Soviet-era town of Chersky. This town of snow and apartment blocks lies 400 kilometres north of the Arctic Circle, where the Sun doesn't rise for almost six weeks each year. It was once a hub for gold miners and scientists heading for research bases near the North Pole but most people abandoned it when the Soviet Union collapsed. Not Sergey Zimov – he was committed to his 'ecological engineering'. He now views Pleistocene Park as an experiment which could ultimately be rolled out across continents and prevent a feedback loop leading to runaway global warming.

Sergey's work began with a failed attempt: he saw no reason why horses would choose to leave his land, and so he let his first herd loose. When they wandered away, he erected a fence, and introduced twenty-five new horses, watching the transformation begin. He has since enclosed 20 square kilometres and has gradually built up the collection of animals. There are now over 150 large herbivores in the expanding park, including moose, musk ox, reindeer and yaks. Some of the animals have been provided by local hunters; others have been brought in by Sergey and Nikita. The pair survived 10,000 kilometres by road and river barge to transport ten yaks, and collected others from the Mongolian border using a military truck. The musk ox came from Wrangel Island in the Arctic Ocean, where the average summer temperature is just 3 °C. The bison were bought from Ditlevsdal Bison Farm, which is just outside the Danish city of Odense, after years of searching for the right animals.

Sergey started his collection of herbivores in 1988, three years before the Soviet Union collapsed and two years before the term 'rewilding' first appeared in print. As rewilding has grown in popularity, Sergey's bold idea has become more mainstream. His strategy has been named 'Pleistocene rewilding', and interest isn't limited to Siberia. Just as Eurasia was once alive with large herbivores, America was home to species such as giant beavers, giant sloths and giant tortoises. Even though these are long extinct, there are other species which could be brought in to play their roles. Inevitably, Pleistocene rewilding has attracted criticism, and some evolutionary biologists believe it is only a slightly less bizarre proposition than *Jurassic Park*.

Even if Sergey and Nikita achieve their ambition, they won't be bringing back the past and returning nature to a pre-human 'pristine' state. Pleistocene Park reveals the futility of attempts to make an accurate replica of a past ecosystem. Not only are they using substitutes for extinct species, they are even using animals created by humans. Their horses are a rare breed adapted to the harsh conditions, believed to have been bred from the domestic horses brought to the region by thirteenth-century immigrants. Although wild horses were once found in Siberia, the Zimovs' Yakutian horses are not their direct descendants.

Pleistocene Park is more representation than a replication of the ice age ecosystem. But what matters to the Zimovs is the roles which animals perform, not the species that are performing them. To them, the Yakutian horse is no less valuable because it is domestic. This is an unusual outlook, as conservation has often been preoccupied with species. Perhaps it is because species are the easiest way we have to understand and categorise nature, or because the beauty and diversity of species is an easy concept to appreciate. However, conservation could just as logically choose genes as a focus, or individual stretches of DNA. Whatever level we look at, from genes to whole landscapes, one thing is clear:

conservation cannot aspire to protect 'pristine nature'. It's not even clear what that is.

Ideas about pristine nature invoke the state that nature was in before humans affected it. The trouble is that humans have played a role in shaping nature for roughly 2.5 million years. Initially this was our sister species, including *Homo erectus*, and for the last 200,000 years it has been *Homo sapiens*. For half that time, our ancestors were confined to Africa, but this changed towards the end of the Pleistocene as populations moved north into the Middle East and on to Asia and Europe. They first entered America from the north Asian mammoth steppe steppe more than 20,000 years ago, via the land bridge between north eastern Siberia and western Alaska.

When *Homo sapiens* arrived in Europe 45,000 years ago, they found habitats already modified by Neanderthals. From the start our ancestors were contributing to extinctions, with Neanderthals an early casualty (although some of their genes live on in modern humans). Humans continued to alter landscapes and witnessed huge changes as the glacial period came to an end and the climate warmed. The simultaneous impact of humans and a changing climate means there is no way of knowing what nature would look like if humans had never evolved; we entered an area that was changing. However, it's clear that our changes were radical. The heather moorlands which appear on Scottish whisky bottles, the Great Plains of the US, and the meadows of the Pyrenees are all human creations.

Even if we relax our view of 'pristine' and look for an absence of contemporary humans, this is hard to find. Plastic has been seen on expeditions to the deep ocean and climate change is leaving its mark on the polar regions. Arguably, nowhere in the world has been left untouched. We therefore often settle for a more recent state of nature. At a first glance this is appealing – we can have more realistic objectives, such as to stop the decline in cuckoos, or bring back the turtle doves. This outlook has become particularly popular for birds: in the UK, declines are reported relative to the 1970s simply because that's when accurate records began. We can certainly choose a recent snapshot in time and attempt to return nature to the state it was in, but we risk having our conservation objective as 'bring back the world I grew up in'. This is no more logical than recreating the ice age.

Settling for a form of nature which is free from recent human impacts also highlights the problem of defining a particular state of nature as 'natural' or 'pristine'. If we are happy to allow the effects of early *Homo sapiens,* then why should we worry about more recent ones? Humans evolved as part of nature – it is not 'unnatural' for us to have an impact on wildlife. Seeing a divide between humans and wildlife is a relatively modern outlook, and is alien to many Indigenous societies. Some native languages don't even have a word for 'wild'. We are all part of nature, and this calls into question any conservation objectives that are based on an ideological separation of humans and wildlife.

There may be no intrinsic reason why habitats modified by humans are inferior, but we can all agree that some of those modifications are negative – when fish stocks have declined and soils have lost their fertility, for example, and that relatively untouched ecosystems such as rainforests bring great benefits, including climate regulation. But it's true, too, that some human impacts have benefited wildlife. The switch from a hunter-gatherer lifestyle to agriculture started a process which would allow billions of people to eat a nutritious diet, and irrecoverably altered life on Earth. The first signs of agriculture date from around 12,000 years ago, and initially were confined to the Middle East. Agriculture spread slowly, but as the human population began to grow the impacts rapidly increased. In some countries the effects were dramatic – by 1350, only a tenth of England remained wooded, a similar area to today. People lament the loss of woodland, and talk about restoring the country to its previous wooded state, yet the early days of farming allowed many species that had previously been rare to flourish. The open landscape provided perfect habitat for ground-nesting birds such as the skylark and partridge, and many species of plant which thrive in grassland benefited, too. There are always winners and losers.

Even if we wanted to, there is no way we can reverse these changes. If we return England to woodland we won't be recreating the past. Deer were much less common in ancient Britain and instead there were herbivores such as wild cattle, which are now extinct. Herbivore populations were controlled by large predators that can no longer be found in the British Isles. The most important tree was the small-leaved lime, with its sweet-smelling flowers providing nectar for a plethora of insects. It is still found in many parts of England, but is now uncommon. Oak came to dominate woodland, because its value for constructing ships and timber-framed houses led humans to nurture it. What we think of as ancient woodland may indeed be ancient, but that doesn't mean it is unaffected by humans. Epping Forest, Highgate Wood and the Forest of Dean are ancient but not 'pristine'.

For Sergey and Nikita, their motivation for recreating this long-lost ecosystem is not because they have a romantic view of what Siberia should look like. Their active ecological engineering would be pretty hard to justify based on an ideology that nature should be separate from humans – as I mentioned, they even bought a 12-tonne tank, hoping it would be an efficient way to clear trees (it wasn't, but it was fun to try). Instead, the seeds of the idea were sown when Sergey was a research scientist. Along with his colleagues, he came up with an estimate of the carbon trapped in permafrost in the northern hemisphere that was double what had been previously thought. Combined with the knowledge that permafrost is warming, this figure becomes terrifying. When permafrost melts, soil microbes transform organic material into carbon dioxide and methane, releasing greenhouse gases. To Nikita, this isn't a threat that can be ignored: 'We are sitting on an enormously big carbon bomb. So, the idea is to find a way to leave this reserve of carbon intact.'

Siberia will suffer deeply if permafrost is lost: the ground will collapse, bringing roads and houses with it, and fish populations will be destroyed when the rivers become mud flows. The Zimovs' main concern, however, is the global impact. If we continue to release greenhouse gasses at the current rate, nothing can prevent the warming which melts the permafrost. However, we can reduce the rate of melting by changing the conditions. Currently a layer of snow insulates the ground – the heat that it absorbs over the summer is trapped over the winter rather than being released into the atmosphere. Nikita explains the options as he sees them: 'How can you cool the permafrost more? You remove the insulation layer. How can you do that? You can either take a bulldozer and bulldoze the entire Arctic, or you can let the animals do that themselves.'

Herbivores need to eat throughout the winter, so they excavate the snow to access the vegetation, which removes the insulation blanket and allows heat to escape. This process of cooling is why Sergey and Nikita are designing a new ecosystem – their objective is to create a better future, not to return to the past. Accusations that they are playing God by engineering nature don't concern them. Humans have always done that, and always will. At least, when they create a new nature, it will bring global benefits. 'Before you protect something, you have to create something,' Nikita says. 'There is no wild nature.'

The Zimovs' vision is for this rewilding experiment to be replicated so that huge expanses of permafrost can be protected. For their own Pleistocene Park, they are still building up the number of large herbivores, and once there are enough it will be time to introduce some predators. However, the animal Sergey most covets is a mammoth, and he is convinced that this will be possible in just a few years. He even has some Harvard University scientists on the case, who have recovered DNA from a 42,000-year-old mammoth preserved in the Arctic ice. In 2015, they inserted mammoth DNA into an elephant genome (rather than cloning a mammoth, they plan to create a mammoth–elephant hybrid). It's a possibility which Nikita is ready to embrace: 'If someone says to me, "I will give you a mammoth next week," I will say, "Yes thank you." And I will prepare for that. If it is a furry elephant, that's good enough.'

The ocean, too, has been irreversibly modified by humans, and changes have brought both opportunities and problems, just as on land. We don't know when humans first realised that the ocean doesn't necessarily bring the greatest benefits when left unaltered, but we know this happened thousands of years ago. Among the most striking examples are the rock walls and terraces of British Columbia. Indigenous people rolled rocks down the beach to build walls at the low tide mark, with flattened terraces behind that were constantly being submerged then exposed by the changing tide. These 'clam gardens' were effective at increasing clam production, although the people who tended them would have needed a long-term vision. The walls were built over the course of generations, with benefits slowly building until clam productivity was increased fourfold. This is a sobering thought – whereas we risk leaving future generations with a depleted ocean, they provided their descendants with a habitat that was becoming ever more productive.

Today, human activities often lead to short-term gains but long-term destruction. Even in the deep ocean, ecosystems can be affected by activities such as bottom trawling, pipe laying and extraction of oil and gas. The victims often include species such as cold-water corals that are long-lived and slow to recover. Unlike many of their warm-water counterparts, cold-water corals in the deep sea don't have symbiotic algae which capture energy from the sun – it is too dark for that. They do, however, form reef frameworks that can last for millennia and act as diverse habitats. They are important refuges and feeding grounds for a spectacular range of organisms, including commercially important fish. We damage them at our peril.

As well as destroying habitats, human activities can also create them. This has led to decades of debate about human structures under the ocean, which range from oil rigs to artificial shelters for lobster fishing. The extent of these structures is staggering. A recent study estimated that human-made structures on the sea floor in coastal regions alone occupy a total area approximately the size of Italy. One big source is shipwrecks, and there are an estimated 3 million ships on the ocean floor. Some of these are thousands of years old, whereas others are recent and intentional, with decommissioned ships being sunk to create new habitats. Many have become popular dive sites, including the *SS Thistlegorm*, which rests at the bottom of the Red Sea. This British Navy vessel sunk in 1941, just months after she set out for her maiden voyage. When anchored off the coast of Egypt, she was hit with two 2.5-tonne bombs and sank along with cargo ranging from wellington boots to two steam locomotives.

Although the sinking was a tragedy for the nine men who lost their lives, the same can't be said for the marine environment. Motorbikes, cars and guns have become encrusted with marine life, and the water around them is alive with fish and turtles; the floor, home to lobsters, shrimps and crabs. This isn't an isolated incident, and it is common for shipwrecks to become home to a spectacular diversity of creatures, including many species that are normally found in caves or dark overhangs. Eels, for example, take shelter in the darkened crevices provided by ship and cargo. Other structures, too, have given rise to new ecosystems – oil and gas platforms, for example, can become highly productive sites for marine life.

The accidental impact of such structures has prompted people to create human-made reefs with the express intention of stimulating a new ecosystem. These attempts haven't always been successful, with blocks of tyres, for example, disintegrating and drifting off to pollute the surrounding sea. We've learnt from mistakes, though, and artificial reefs are becoming increasingly innovative. Some metal structures are even electrified, and the low voltage current protects them from corrosion and promotes mineral deposition. Corals have been found to grow more quickly on these structures and to have increased resilience to high temperatures and other disturbances.

Human structures can be things of beauty, and the artworks of underwater sculptor Jason deCaires Taylor provide striking examples. He is a prolific artist, with creations installed around the world. In 2006, he constructed the world's first underwater sculpture park, which *National Geographic* declared one of the twenty-five wonders of the world. His statues often represent the actions of people living above the waves, oblivious to the damage they are doing to the natural world. One statue in Mexico, *The Banker*, depicts a man in a suit with his head buried in the sand; it was created in the wake of the 2008 financial crisis, and has since been joined by a collection of figures with briefcases. In 2016, deCaires Taylor created Europe's first sculpture park in Lanzarote, a collection of 300 submerged sculptures and architectural forms.

As well as raising awareness about the plight of the marine environment, the sculptures are designed to bring direct benefits. Each one is created using non-toxic, pH-neutral marine-grade cement, with a rough texture that encourages coral larvae to attach. They are shaped to provide homes for fish and crustaceans, including the folds in the figures' clothing. Gradually the artworks become colonised by the creatures of the sea, taking on a diversity of shapes and colours that human artists would never have imagined. Coraline algae covers the sculptures in shades of purple, and pink sponges grow across the surface to give the impression of skin laced with veins. Fans of soft coral hang off sculptures like wings, while staghorn coral grows like strange antlers. Eventually, the original shape of the statue is lost.

DeCaires Taylor is continually adapting his statues to create better underwater habitats. When he found that lobsters living in his sculptures had been caught by fishers, he made a hollow statue of a VW Beetle with holes in the windows to give lobsters access to a safe chamber inside. He depicted a crying child lying on the bonnet to symbolise what we are leaving for future generations, and huddled against the windscreen to hide her tears.

Not only do these sculpture parks create new ecosystems, but they also take the pressure off existing reefs by drawing tourists away from them. They raise funds for marine conservation, provide employment for local people, and help build momentum for protection. DeCaires Taylor's first park, situated in the Caribbean Sea off the coast of Grenada, was instrumental in the government declaring the site a marine protected area (a topic we will revisit in Chapter 7).

There can clearly be many benefits from human-made structures on the seabed, but their great diversity means it's hard to generalise about their impact. Criticisms include a fear that we are favouring the colourful species that benefit from these hard structures rather than the hidden life found in the soft sediment of the sea floor. Some people view human structures as careless pollution, and believe they conflict with an outlook of respect for the natural world. However, there's no intrinsic reason why the ocean is inferior if it contains human structures, and no reason that the novel living communities that inhabit them are less valuable than those found on natural reefs.

New reefs are often home to a unique assemblage of species, which is unsurprising given that the physical structures are different to those created in nature. They provide a unique collection of niches, and can bring chemical changes, too. For example, shipwrecks can have high levels of sulphur from cargo such as gunpowder, or sulphides can be released by bacteria living on the rotting wood of the ship's hull. This attracts organisms such as tube worms, which are adapted to survive in deep-sea hydrothermal vents, again contributing to a unique ecosystem. Even organisms themselves may be changing, sometimes as a by-product of wider changes, and sometimes as a result of concerted efforts. Most notably, corals are being bred to resist higher temperatures.

To embrace these changes, we need to let go of a separation between humans and nature and a desire to recreate a past baseline. As marine biologist Dr Sofia Castelló y Tickell from the University of Oxford says: 'If we think of human-made reefs as imitations of natural coral or rocky reefs, this can make us blind to the unique benefits they may bring. In a changing ocean, we can't afford to miss the conservation opportunities provided by the structures that humans create.'

Even in habitats that are extensively modified by humans, nature can thrive and bring great benefits. Bringing more nature into Augustenborg, a district of Malmö on the southern tip of Sweden, reversed its fortunes at the turn of the millennium. In the 1980s and 1990s, the area faced social and economic decline and suffered from regular floods when the sewers couldn't cope with heavy rainfall. One solution would have been to increase the capacity of the traditional sewage system, but a cheaper option was presented by nature. Ditches, ponds, wetlands and green roofs were created to absorb the excess rainfall, and these benefited many species of wildlife.

Ponds make amazing wildlife habitats, as the newts, frogs, snails and many insect larvae which quickly colonised my garden pond can testify, and insects and birds were attracted to Augustenborg's green roofs. The image of the district was reversed from one characterised by social decline and overflowing sewers to a neighbourhood of green space. The city had created a new environment where humans and nature can thrive.

The thriving of urban wildlife raises an interesting question for conservation. What do we mean by 'wild'? Not only is an urban habitat so very different to what went before, but individual animals and plants become interconnected with people. The frogs in Augustenborg rely completely on ponds created by humans. This is a common story, as any walk in a garden or park will show. Where do you draw the line between what is domestic and what is wild? Rose bushes, we would probably agree, aren't wild. They are planted by people and their genetics have been altered by breeding. The bumblebees which feed on them, on the other hand, fit most people's definition of wild. The bees you see on roses are almost certainly from colonies which aren't managed by humans, even though they rely on the plants we tend. Few people would exclude them from a definition of wild. If we went down that route, we would have to exclude a blue tit which uses a nest box or a bat which roosts in a roof.

If you are in an urban park the chances are that you will see a feral pigeon, which also raises issues. The pigeons we see around the world are descended from rock doves, which were once confined to the cliff ledges of southern Europe, North Africa and the Middle East. Wild rock doves were tamed and bred to be pets and messengers, and have even played an important role in wars. A British pigeon trained by American pigeon fanciers, Cher Ami, famously saved lives during the First World War. When a trapped battalion came under friendly fire on the Western Front, they released Cher Ami with a message describing their location. She was shot down almost as soon as she was released, but managed to take to the air again and made it back to her loft with the message. Despite her injuries, army medics saved her life. She had lost a leg, but she received a wooden replacement along with a Croix de Guerre medal.

Although she could fly wherever she chose, we would probably call Cher Ami a domestic animal, as her fate was intertwined with that of her human handlers. But what about her relatives living free in the world's cities? They are definitely not pets, but they rely on food provided by people and are restricted to human environments. If we decide that pigeons aren't wildlife, then what does that mean for other urban birds? Peregrine falcons have also made their home in cities, nesting on buildings and bridges – and feeding on pigeons. In New York City, pigeons also provide food for a dynasty of red-tailed hawks founded by a bird known as Pale Male in 1991. When this male bird with an unusually pale head took up residence in Central Park, he became one of the first red-tailed hawks to nest on a building, and has since had at least eight partners. Red-tailed hawks now commonly nest around New York.

One difference between urban hawks and feral pigeons is that pigeons have had their genetics altered by people. Domestic pigeons have such varied genetics that Darwin bred them in his garden to gain a better understanding of evolution in the wild. Feral pigeons may not have quite such elaborate plumage as the ornamental breeds that Darwin favoured, but they do show striking physical and genetic differences to their wild ancestors.

This difference is very obvious in pigeons, but subtler differences are widespread. Humans are altering the genetics of a vast number of species, many of which are firmly in the category of 'wild'. For example, extensive fishing has driven evolution in Atlantic cod so they now mature earlier. Feeding wildlife has also produced genetic changes. In Britain, the constant supply of food from bird tables and feeders has changed the migratory route of blackcaps. These tiny birds have increasingly been spending the winter in

British gardens rather than migrating to Spain. Anyone lucky enough to see blackcaps on their garden feeders over the winter is both creating and witnessing rapid evolution.

Returning to those roses, the definition of wild can be equally murky for plants. Dandelions are generally seen as wild – they may be common in gardens because we have created ideal conditions for them, but they spread by themselves and aren't genetically altered by human breeding. I must be one of the very few people who actively cultivates dandelions, nurturing them as food for the guinea pigs. Are they wild, while the spinach growing next to them isn't? What about if I'd scattered dandelion seeds on my lawn, or sown a bag of wild flower seed mix.

Wild flower seeds provide a particular contradiction. 'Wild' is in the name because they are species which haven't been genetically altered by humans, yet these individual seeds are planted by people. Can plants which start life in a packet be considered wild? In reality, there are few plants and animals which aren't affected by people – from the giraffes in African game parks to the oak trees planted in British woodlands. It's therefore unsurprising that 'wild' isn't a clear category. Just as ecosystems are hugely affected by people, so are individual animals and plants. And, like ecosystems, individual organisms are not automatically inferior if humans have affected them. My garden hedgehog is not tainted because I fed it peanuts, and nor are peppered moths because urban pollution changed their genetics to make them camouflaged on darkened tree trunks.

In most of these ambiguities, humans have gradually increased their impact on a species. However, there are also striking cases where their influence is being reversed. One of the most controversial is Australia's dingo, whose origins and status have been hotly debated. The oldest known dingo fossil dates back to 3,500 years ago, and they probably arrived in Australia not long before that. They were brought by humans, as domestic dogs which had long since diverged from their wild ancestor, the wolf. Details of what happened next are shaky but, by the time Europeans arrived, dingoes were widespread as wild animals. They had become distinct from other wild dogs – even their close relative the New Guinea singing dog – but still bore a resemblance to many dogs living wild in Asia. The European settlers brought domestic dogs to Australia for a second time, and the dingo's story became even more complex. They now regularly hybridise with domestic dogs, and it can be hard to tell a pure-bred dingo from a hybrid.

This history has led to differences in opinion about how to classify dingoes and how they should be treated. Are they a subspecies of the wolf? A breed of domestic dog? Or a species in their own right? For some, this definition is important, as is the imperative to preserve dingoes exactly as they are. There have been calls to shoot animals which are visible hybrids between modern dogs and dingoes, to try and prevent dingo genetics from being polluted.

Others care little for dingoes' genetic 'purity' or historic origin, and instead focus on their ecological impact. But even this brings a controversial debate. Dingoes can be seen as a threat to the native animals they prey on, or as providing the valuable service of controlling pests such as foxes, rats and bandicoots. Lack of data makes it easy to fit the evidence around intuitive beliefs, which can be tied up with a fear that they will catch livestock (although it's clear that feral cats and dogs kill more native prey than dingoes). The uncertainties and polarised debate mean that the dingo has an ambiguous legal status, protected in some areas of Australia yet classed as a pest in others. Money is simultaneously spent on their conservation and their eradication.

Other domestic animals have returned to living wild or semi-wild more recently, and some are even being used for conservation. The primitive Konik pony from Poland, for example, is a popular choice for maintaining grazed vegetation, as is the Highland cow. Some people have wanted to take this a step further and create breeds more akin to extinct wild ancestors. Among the most famous examples are Heck horses and Heck cattle, created by Heinz and Lutz Heck. These two brothers grew up in the grounds of Berlin Zoo, where their father was director, and both went on to become directors of German zoos. This left them with the skills, contacts and resources to follow their passion for animal breeding. While their contemporaries in the 1920s tried to create ligers and tigons, the Hecks turned their attention to trying to recreate the auroch, Europe's wild cow.

Aurochs are the ancestors of domestic cattle, and they went extinct in 1627. We know from their skulls that they had formidable horns, similar to those of fighting bulls, and that they were larger than their domestic descendants. The males are thought to have been black with a pale stripe running down their back, and stood at a mighty 1.8 metres. Females were somewhat smaller and their coats were the rich brown of a chocolate Labrador. Further details are hazy.

Although aurochs had been extinct for hundreds of years, Heinz and Lutz believed that selective breeding of modern cattle could bring out their ancient features. It was just a case of ridding them of weaker genes that arose during the process of domestication. The brothers were inspired by prehistoric cave paintings and historical accounts that often described the animals' ferocity. Temperament was a particularly important issue for Lutz, who was keen to seek out aggressive animals and introduced fighting cattle from Spain into the breeding stock. In 1939, just twelve years after the breeding project had begun, he proclaimed 'the extinct Aurochs have arisen again as German wild species in the Third Reich'.

Lutz introduced Heck cattle to the occupied Białowieza Forest, which now straddles the border between Poland and Belarus. Here, their true wildness, as he saw it, could be realised, free to do battle with human hunters. However, the demand for agricultural land meant they weren't introduced as widely as he had hoped, and his forest herd was slaughtered at the end of the Second World War, leaving just small numbers alive under Heinz's care in Munich Zoo.

This could have been the end of the story, but interest in the cattle was rekindled in the 1980s, when they were exported to the Netherlands for a pioneering rewilding project at Oostvaardersplassen (which we will return to in Chapter 5). Here they fulfilled their ecological role as grazers, with the advantage that they needed little human management and could withstand cold winters on poor ground.

In recent years, other research projects have tried to create something closer to the auroch – and a little less aggressive than the Heck herds. One such is the Tauros programme, which employs a professional artist as a core part of the project team to visualise archaeological materials into depictions of aurochs that can guide the project. Scientific advances are used alongside the artistic visions, and a milestone came in 2015, when the auroch's genome was sequenced from a single fossil. This made it possible to select modern breeds with the closest DNA sequence to the auroch and interbreed them using artificial insemination. By the end of 2017, there were over 600 Tauros cattle, and the project began to transition from domestic to feral populations. The intensive breeding methods of the project's early stages could be replaced by natural selection.

Meanwhile, other scientists have been attracted to the idea of using new advances in genome-editing technology to create a beast whose DNA is an even closer match to its ancestors. The idea is that this genetic purity will bring us closer to a true auroch, not a second-rate copy produced by breeding programmes. But, if cutting-edge technology produces a cow with genetic codes matching its ancestors, is that more or less natural than a lookalike created through selective breeding? What is most important – the cow's looks, behaviour, ecology or genes? Not everyone will give the same answers to these questions, and the different outlooks reveal how 'natural' is a cultural concept.

Heck cattle certainly have origins based on a culture we now reject. As an active member of the Nazi Party, Lutz Heck's vision for the auroch was tied up with an ideology in which rural life was romanticised and the urban environment was viewed as a site of moral degeneration. Nazi propaganda presented Heck cattle as helping reconnect elite, alienated Germans with their true natures. In the 1940s, the discourse around care of the landscape was used to justify the displacement and purging of Eastern Europe's current human occupiers. The cattle's cultural context has now shifted, but they are still part of a narrative of reconnecting people with wilder, more wholesome landscapes and counteracting 'nature-deficit disorder' among city dwellers.

For some people, the image of Heck cattle is too entwined with their Fascist origins. When a farmer in Devon imported a small herd with the ambition of running safaris on his land, Britain's tabloid press delighted in reporting tales of 'cattle created by Hitler'. In the end, he had to cull half of the herd due to their aggressive behaviour.

Of course, using Heck cows in rewilding projects in no way endorses the political outlook which surrounded their creation. However, Lutz's desire to return Eastern Europe to a purer past does serve as a reminder of where our conceptions of wild can lead us. It's not an isolated incident, and colonial societies have too often been guilty of dispossessing people based on an imaginary idea of purity.

Wilderness often exists in our heads as an imagined ideal of purity, but this outlook of a separation between nature and humanity is a recent cultural perspective. Whereas Indigenous people often have lives that are interwoven with nature, our very different perception has been shaped by the outlook of our ancestors. This is particularly clear in the USA.

In the late nineteenth century, preserving nature in its so-called 'natural' state became the bedrock of American environmentalism. There was an aspiration for 'true nature', separate from people, and so America's first national park, Yellowstone, was sold to tourists as a pristine and uninhabited wilderness. The writings of Theodore Roosevelt give a fascinating glimpse into how early pioneers of wilderness preservation saw the world. His book *Hunting the Grizzly and Other Sketches* opens with the assertion: 'When we became a nation in 1776, the buffaloes, the first animals to vanish when the wilderness is settled, roved to the crests of the mountains which mark the western boundaries of Pennsylvania, Virginia, and the Carolinas.'

To Roosevelt, America was what came after European settlement, and wilderness was what was there before. It is clear that he romanticised both those times; he loved his nation as well as the landscape it was eroding. His writing also reveals that his view of the conquest and protection of wilderness was tied up in a masculine identity. He saw white hunters and trappers as 'dauntless and archetypical wanderers', and believed his family displayed these noble characteristics. The book's opening chapter describes a hunting expedition in which his brother Elliott, his cousin John and their fellow 'adventurers' encountered many thousands of buffalo. The party survived hunger, thirst, a buffalo stampede and a visit from a 'war party' of Native Americans. Roosevelt describes the hardship of their exploits, which he concludes to have been a 'very pleasant and exciting experience'. His vivid writing glories in the rugged life of freedom and conquest.

But the buffalo were disappearing. Over 60 million animals had roamed America in vast herds in the late-eighteenth century, yet fewer than 600 were alive in 1889. This fact wasn't lost on Roosevelt when he rose to the presidency two years later. In many ways he was ahead of his time. While his contemporaries still considered natural resources to be inexhaustible, his thoughts turned to what would happen once they were gone. His achievements include creating five national parks, four national game preserves and 150 national forests. He made fundamental contributions to the development of wilderness preservation legislation, and this strategy continues to shape American conservation.

But again, the preservationist attitudes had a darker side. The idea of 'pristine' parks only works if there is no trace of people, and so evidence of Native American culture was removed, often along with the Native Americans themselves. The view of wilderness as 'true nature' set apart from humans made it acceptable to remove people from their territories. Land that had been shaped by millennia of Native American culture was reframed as virgin wilderness and their history was erased. This outlook was clear in the writings of the early American conservationists, many of whom also express their love of hunting. Critics therefore point out that the early national park movement was a convenient way for the rich to secure their hunting rights. Such 'green land grabs' can be seen as colonialism in a different guise, and the threat continues.

Dispensing with the idea that there is an objective 'natural' state of nature opens up huge possibilities for what conservation should look like and what it can achieve. Once we are free to celebrate nature in all its forms, there is much greater potential to ensure there is more of it. We can move on from futile attempts to restore nature to the way it was, and instead help it thrive in new ways. We can recognise the value of nature in city parks, not just primary rainforest. Perhaps most importantly of all, we are freed from seeing humans as the enemies of nature. Nature and humans are entwined, and we need to embrace this, not try to separate ourselves from the natural world which sustains us.

Conservation is about the future, not the past. Instead of returning nature to an 'ideal' state, we are free to create a new landscape which meets the needs of humans and animals, now and in the future. This is certainly how Sergey sees Pleistocene Park: 'Wild ecosystems were our enemy for all our history. It's possible to turn wild nature from our enemy to our friend, but we must begin this now. We'll have victory and mammoths will own this territory.'

Unlike our ancestors, we have access to vast amounts of knowledge, and we understand the impact of our actions. Combined with our incredible ability to change nature, this gives us the power to create the future we want to see. But what future do we want for wildlife? And why? The Zimovs are a rare example of conservationists with a clear reason for changing nature: they want to prevent a feedback loop of permafrost melt that would dramatically increase global warming. A recent study estimates that carbon emissions from thawing permafrost could be around 4.35 billion metric tonnes per year over this century – around half as much as from fossil fuels. It is their powerful answers to questions of 'why' that have prompted strangers to donate tens of thousands of dollars to the unlikely crowdfunding campaign of transporting bison the length of Siberia.

The challenges associated with turning vast Arctic areas into a Pleistocene Park are considerable, and we would need a much larger area to determine if it brings the benefits they hope for. However, massive problems such as climate change are going to need creative solutions, and experiments such as this one help us think in new ways. Our eyes are being opened to possibilities such as introducing species into new areas, rethinking attitudes to domestic animals and embracing and conserving 'modern' and urban landscapes.

For Nikita, the imperative for doing this is clear: 'I have three kids, and what I do, I do for them.'

CHAPTER 2

THE BEAUTIES AND THE BEASTS

Bees vs wasps (and the joys of parasites, microbes and marine sediment)

No visit to Minnesota's Twin Cities is complete without a trip to the Mall of America – or that's what my colleagues at a climate conference in Minneapolis told me. So, when I had time to spare on my way to the airport, I accepted the irony of stopping off at a monument to consumerism before a long-haul flight and entered the largest mall in the USA. My colleagues weren't wrong about it being a sight to behold. The Mall is a giant windowless world, echoing to the whoosh of rollercoasters. What ended up drawing my attention, though, was a stand run by The Nature Conservancy, the largest environmental charity in the Americas.

Perched in the fluorescent light of a shopping aisle, this was little more than a few boards filled with Post-it notes. But below the boards were piles of pens so that passers-by could add their thoughts about protecting nature. It gave visitors a chance to reflect before buying a new pair of trainers, and their reflections were fascinating. Alongside sentiments such as 'Save the Ocean' and 'I ♥ Roses' was the repeated message of 'Save the Bees!!' The message came over loud and clear: this is where we expect conservation to focus.

Nowhere on the board was a mention of wasps. There never is. It's a fact that Professor Seirian Sumner is on a mission to change. An entomologist at University College London, Seirian is co-founder of the Big Wasp Survey (and the Twitter hashtag #WaspLove), which was set up to raise awareness of wasps in the UK and collect valuable data about their distribution. She certainly has her work cut out to change the wasp's image: 'When I meet somebody and tell them what I do for a living, the most common response is "What's the point of wasps?" Most people hate them.'

Wasp species come in a tremendous variety of sizes, colours and lifestyles, ranging from the Asian giant hornet with its 7.5-centimetre wingspan to fairy wasps measuring less than one millimetre. But much of this staggering diversity is unknown to science: there are approximately 110,000 known species, but probably hundreds of thousands left to discover. Most wasp species are solitary. However, around 1,000 species live in social colonies and these are Seirian's focus, in particular yellowjacket wasps. These yellow and black insects – the quintessential wasp – are a small group of species but with a wide distribution.

When Europeans consider wasps, we are normally thinking of the common wasp, or the almost indistinguishable German wasp. In North America, a number of related species go by the name of 'yellowjacket'. These common species play important roles in our ecosystems, as pest controllers, and in providing food for a range of other wildlife. As Seirian points out: 'In the UK, badgers are one of the key predators of wasp colonies. They are unaffected by the stings, so just dig into the nest to get to the larvae, which are incredibly nutritious. Other predators include birds, spiders and other wasps, and in the Americas yellowjackets are food for racoons and coatis.'

Wasps in turn provide amazing pest control services, helping to regulate populations of other invertebrates. This service is vital, given that insects breed at a rate which puts rabbits to shame. A single aphid could produce a billion descendants in one season if they were all to survive. Aphids can even be born pregnant: females can produce clones of themselves without needing to mate, and give birth to nymphs that already have embryos developing inside them. If this breeding continued unchecked, in just one year the UK would be covered in a blanket of aphids 150 kilometres deep.

Thankfully, they don't all survive, and wasps are part of the reason. In the UK alone, it is estimated that wasps consume 14 million kilograms of insect prey – including caterpillars, flies and aphids – each summer. That's the combined weight of almost 3,000 elephants. As Seirian says: 'In a world without wasps, these other species would become a problem. Even now, we try to control insects with chemical pesticides, but we know that excess use is having a detrimental effect on other wildlife. We need to look to nature's solutions instead.'

As well as preying on other insects, wasps drink nectar, hence they terrorise autumn picnics in search of sugary substitutes. This is a misfortune for picnickers, but in fact brings great benefits – when wasps visit flowers in search of nectar; they play an important role in pollination.

These twin services of pest control and pollination arise because of the wasps' complex societies. Adult worker wasps drink nectar and catch prey to feed their growing sisters. Just as honeybee colonies have a single queen and thousands of workers to rear her offspring, so do yellowjacket wasps. The colony work together, each performing their assigned role. Our romantic views of honeybees behaving as 'one for all and all for one' could just as well apply to yellowjackets.

The behaviour of social wasps attracted Seirian to study them for her PhD. Wasps, she discovered, share the characteristics that attract us to bees, as pollinators and as individuals working tirelessly in service of the nest. A few wasp species even produce honey. But Internet searches for 'wasp pest control' tell you how to get rid of them, and news stories about wasps tend to be along the lines of 'residents barricaded in homes by plagues of wasps'. High wasp numbers in 2018 led to headlines declaring: 'Wasp numbers are worst on record'. An increase in bee numbers, on the other hand, would be a cause for celebration. As Seirian notes: 'We care about bees because we understand that they pollinate our crops and the beautiful wild flowers. If a bee is caught in your house you don't get the insect killer out, you carefully open a window and spend ages trying to get it out.'

Such cultural attitudes to wasps and bees have an ancient history. Aristotle wrote that wasps lack the extraordinary features of bees and biblical punishments include plagues of hornets (a group of wasp species closely related to the yellowjackets). Honeybees, by contrast, are part of our human cultural identity. Archaeological finds date beekeeping back at least 4,500 years, so it is not surprising that this amazing symbiosis between bees and humans is embedded in our psyche. The connection runs deep in many cultures – in Hungarian, for example, the word for 'bee' also means 'womb', and the word to describe the death of a bee is the same as the death of a human, not the death of other animals.

This cultural history feeds into modern conservation priorities. Intuition guides us towards a focus on bees, while wasps are ignored or persecuted. But we need to question whether this make sense. With limited funds to go around, we can't afford for a few beautiful species to steal all the limelight. For all the money spent on pandas, rhinos and honeybees, hundreds of thousands of species receive nothing. In the case of wasps, intuition has clearly failed us: these wonderful creatures don't get the respect and protection they deserve. But has our intuition served us well in attracting us to bees?

The answer is surprisingly complex, and to address it we need to consider what we mean by bees. We often use 'bee' and 'honeybee' interchangeably, but the Western honeybee is just one of over 20,000 bee species in the world, with 2,000 found in Europe and 5,000 in North America. Even bumblebees are a tiny fraction of the bee diversity, with fewer than 250 species described from around the world. Like wasps, the vast majority of our bees are solitary. Instead of living in large hives or nests, they lay their eggs in holes in the ground or trees, or even in the hollow sticks of my home-made bee hotels. Some are emerald green, some are jet black, and others have the same yellow stripes as honeybees. Most are given no more conservation attention – hardly surprising, given their general neglect by science.

We may not know the detailed trends for many bee or wasp species, but we do know that overall insect numbers are declining in many parts of the world. In the twentieth century, the greatest insect declines were in North America, though more recently declines may have slowed in America and accelerated in Europe. In 2020, scientists produced an analysis of long-term studies from more than forty countries, revealing that insects and spiders on land, on average, had declined by 9 per cent per decade. The situation is alarming, so it's clear that we need to act – and to act wisely.

For millennia, the Western honeybee (*Apis mellifera*) has provided humans with honey, wax, pollination and inspiration. All these offerings relate back to their social lifestyle, in which tens of thousands of sterile female worker bees work together to rear the queen's offspring. Part of this work involves collecting nectar to make honey, which becomes the hive's food supply whenever there isn't enough nectar available. Such a nutritious food source inevitably attracts the attention of other animals, and humans in particular work hard to get hold of the honeybee's winter food supply. Initially this was simply done by raiding nests, but beekeeping has evolved over millennia, as have the bees.

Humans have changed honeybee genetics by breeding bees so they are good honey producers, aren't too aggressive and are resistant to disease. Initially all people could do was select queens from favoured hives, but in recent decades we have taken things to a new level with artificial insemination. Left to their own devices, each queen would mate with multiple males, but only in their first ten days of life. This sperm lasts for the three to five years she will live, laying up to 2,000 eggs per day. This is fortunate for bee breeders: anaesthetising a queen with carbon dioxide, then injecting a syringe of semen into her abdomen under a microscope only needs to be done once to last her a lifetime. To some conservationists, this fact alone means that honeybees are not worthy of conservation attention: they are no longer truly a wild animal. However, as we saw in the previous chapter, the distinction between domestic and wild can be ambiguous and artificial. Although it may be of practical importance, there is no intrinsic reason why 'wild' is superior. There are some ancestral strains of honeybee still living wild, even in Europe, and they attract conservation attention. These wild honeybees have characteristics worth preserving, such as resistance to disease, but their genetics are not automatically more important than their domesticated relatives.

Likewise, some people argue that we shouldn't protect honeybees outside their historic range. Honeybee declines may be lamented in America but they are only present because humans took them there. Again, however, the bees' past distribution may have practical significance, but there is no intrinsic argument for honeybees only 'belonging' in their historic range of Africa and Eurasia.

Another argument against honeybees as a conservation priority is that their populations are thriving (at least those under human care). The United Nations estimates that there are over 90 million beehives in the world, and that figure has been steadily increasing in the last twenty years. Honeybee numbers are growing in the UK, having dipped around the turn of the millennium. These numbers are in stark contrast with media warnings. So how do we explain this discrepancy?

Part of the explanation is hype. But some of the concerns are genuine – while the global trend for honeybees is upwards, that's not true everywhere. Germany, Sweden and the USA have all experienced honeybee declines. Even where numbers aren't falling, some beekeepers report challenges, including hives dying off over the winter. Global trends therefore mask local problems. Conservation needs (and impacts) can differ very widely by location.

Honeybees illustrate this point well. In some situations, they have negative impacts on other wildlife. There's evidence that honeybees compete with other bee species for nectar and pollen, which can become a problem when there aren't many flowers around. They may also spread disease, and recent studies suggest that they can transmit viruses to bumblebees. The scale of these negative effects is hotly debated, with some scientists considering them minor and others calling for hives to be banned from national parks and nature reserves.

However, honeybees also bring benefits, both to humans and to other wildlife. In a landscape modified by agriculture, some wild flowers suffer from a lack of pollinators and honeybees can sometimes compensate. And conservation schemes designed for honeybees may have wider benefits. Creating wild-flower habitats, for example, provides food for many birds and insects, and often raises the profile of conservation. As Seirian says: 'The bee success story has made people much more aware of nature, and the fact that they shouldn't just try to kill everything that crawls or flies.'

Elsewhere in the world, care for honeybees prompts other positive interventions. In Africa, forest beekeepers make simple hives from local materials, then place them in trees and wait for a swarm to move in. Selling the honey can make forest more profitable than farmland, so people are less likely to fell trees to make way for fields. Instead they will care for the forest. In Zambia, for example, beekeepers work hard to prevent bush fires; in Ethiopia, they tend the seedlings of trees that are popular with bees.

Of course, nothing is simple, and conservation aimed at honeybees won't help all wildlife. Honeybees are very adaptable and can benefit from small patches of habitat that won't be suitable for all species. Sowing a general wild-flower mix may be ideal for honeybees, but it won't help insects which need other specific flowers. And the very best ways to increase honeybee numbers don't help other species, because they are all to do with beekeeping. The fortune of honeybees is largely dependent on the skill and dedication of beekeepers, who have many tricks up their sleeves to keep the numbers of honeybees high. They can treat the hives with pesticides to protect the bees from parasitic mites, for example, and if a hive doesn't survive the winter they may split another colony in two to keep up the numbers. These strategies don't benefit other species.

Do honeybees warrant their priority treatment on our metaphorical ark? People often cite pollination as a motivation for protection of bees, but while pollinators play an important role in our food supply, the proportion of our food that requires animal pollination is estimated as 5 to 10 per cent. This figure is lower than most people believe (we could still grow at least nine out of every ten mouthfuls without pollinators). It is still a staggering amount of food, though, and tends to be the nutritious foods we value highly, such as fruit and beans. For this reason alone, pollination is a good argument for honeybee conservation.

However, focusing only on honeybees won't meet our need for pollination. Honeybees share credit with a whole range of pollinators, including flies, moths, wasps and bats. Many wild bee species are more efficient pollinators than honeybees, and they are particularly important for certain crops such as apples. Relying on a single pollinator also leaves us vulnerable should they be affected by disease. So, while focusing on the honeybee is one way to increase pollination, it is not always the best one. Wasp conservation could also help us achieve our goal.

Of course, there is no single motivation for conservation. Some farmers may simply want to increase their production by conserving honeybees. Other people may value honeybees for their roles in pollinating wild flowers and feeding birds, or for the pleasure they (and the birds) bring us. The honeybee's cultural importance is a powerful motivator, too – a world without honeybees would be a sadder place. But let's not be too quick to focus on the beauties; we need to give the beasts a chance, too.

Why, for example, might we want to conserve parasitoid wasps? Or indeed the parasitic hairworm?

If there's one thing that might make a wasp less appealing, it's a lifestyle based on parasitism. Parasites are generally viewed with disgust, yet there is much to be said in favour of the roles that these incredible organisms play. For example, parasitoid wasps kill their hosts and thus perform the same pest control services as yellowjackets.

Globally, there are believed to be around 650,000 (mostly undiscovered) species of parasitic wasp. They are often dismissed as tiny flies, but once you get your eye in you'll find they are all around you. I frequently see them loitering in our vegetable patch where there are plenty of host insects. If you look closely, a distinguishing feature is the ovipositor that females use to lay their eggs in caterpillars or other unsuspecting hosts. This very thin tube extends from the tip of the abdomen, and ranges from a tiny point to a spike that is longer than the wasp's body.

Their role in pest control means that parasitoids are welcome guests on my vegetable patch, but this is just the start. Parasites, in many complex ways, shape the ecosystems we love and rely on, and are important components of the food web. One iconic example is the oxpecker, which uses its bright red beak to pick off ticks and other parasites from animals such as zebras and giraffes.

More often, though, the role of parasites is rather less glamorous – and a lot harder to spot.

Sometimes parasites actively benefit the organisms they infect. Those which cause little harm can protect their hosts against more dangerous parasites. Certain parasitic worms accumulate heavy metals in their bodies, and so protect their host from these toxins. This example raises questions about the definition of 'parasite', given its potential benefit to the host. Part of the ambiguity is that benefits may only occur in certain environments – if no toxins are present, the worm brings only problems. Given the complexity of these interactions, making informed decisions about conservation can feel like an overwhelming challenge. It can be almost impossible to predict the impact of a decline or increase in a particular species, so conservation ignores parasites at its peril.

As the weird story of the kirikuchi char reveals, the effect of parasites can have a surprising reach. The kirikuchi char is an endangered trout, found only in a small area south of Kyoto, Japan, where biologists are making a concerted effort to protect its dwindling populations. The char feed on insect larvae and other invertebrates in the stream, but as the summer draws to a close this changes – there suddenly becomes an abundant supply of crickets that have jumped into the stream from the surrounding forest. These crickets contribute an amazing 60 per cent of the char's annual calories. But why would a cricket take a suicidal leap into a stream? Studies have revealed that this is down to mind control by a parasitic worm.

The worm in question is a species of nematomorph (also known as horsehair worms or hairworms), and has a complex life cycle involving two different hosts. The adults are thin and brown, growing up to 30 centimetres long, and live free in a stream for the final part of their life. The adults don't eat, and spend their energy producing the next generation – an individual worm can lay tw2o million eggs. The larval worms then infect insects such as mayfly larvae that are developing in the stream. Eventually, the adult mayfly flies free from the stream, at which point it has just hours or days to live.

When the dead mayfly falls to the forest floor, it can become food for a hungry cricket, which is exactly what the parasite needs. The hairworm continues to grow and develop in the cricket, absorbing fats and the reproductive organs. Only tissues that are essential to the cricket's survival remain – everything else gets digested by the hairworm. The worm, however, is stranded on the land, with no way of ensuring its offspring can reach aquatic insects to host their development. Hairworms have solved this problem by manipulating the crickets' brains so they become attracted to water and jump into the stream. The hairworm then wriggles free of the cricket, leaving its dying host twitching on the surface, and swims away to find a mate. (If you want to watch this gruesome spectacle, a YouTube search for 'hairworm' will bring up videos of worms squeezing out from their 'zombie' hosts.)

Compared to uninfected crickets, those with hairworms are twenty times more likely to end up in streams. This is good news for char, and potentially for other species too. If the fish, satiated by crickets, eat fewer of the other insects in the stream, this could increase the insect diversity. It's not just the stream that would be affected – many of the insects are larvae, and the adults will fly free into the forest. Hairworms therefore play a vital role in the ecosystem, and connect the land and the stream.

Like so many species, though, this parasite faces problems. In the early twentieth century, conifer plantations replaced many of the forests lining the Japanese riverbanks. These forests are part of a cycle of logging, in which large areas are felled every forty years. Each time the forest is cut down, hairworms are lost from the area, and their population is slow to recover.

This knowledge can help create management strategies to reduce the impact of conifer plantations on the stream, but it poses challenging questions. Should we return the plantation to native forest in order to increase the numbers of hairworms and char? If so, who should foot the bill? From the complex life cycle of a parasite, we quickly arrive at questions about what we want the natural world to look like and what we are willing to sacrifice to achieve that. The hairworm, char and stream are only part of the picture – the way we manage our forests will determine their long-term sustainability. Managing forests and landscapes are topics we will return to in later chapters.

While the repellent aspects of wasps and parasites may be the reason they get little conservation attention, many species are ignored because we barely think of them as part of nature. Conservation rarely considers the microorganisms, yet in recent years there has been a deluge of research revealing how they shape our environments, our bodies, and even our minds. When we talk of protecting the diversity of life, we seldom think of the microbes we can't see. They are, however, the species that make the rest of life possible.

Like the vast majority of microbes, *Mycobacterium vaccae* is known so little beyond the people who study it that it hasn't even been given a common name. Its scientific name comes from the Latin for 'cow' (*vacca*), as it was first isolated from cow dung, and we now know it is related to the tuberculosis bacterium. It is widespread, often to be found in soil and compost, and oddly fascinating.

As *M. vaccae* multiply, they form a layer of cells known as a biofilm. This is invisible to the naked eye, but a microscopic image reveals long filaments of bacteria tangled up together. The cells can share nutrients, and working as a unit makes them more resistant to stressors. Research has shown that in mice the bacteria can activate serotonin in the brain, helping to control stress responses. When it is injected, it alters the animals' behaviour in a way similar to antidepressants. These effects have been found in fish, mice and rats, and the bacteria seem to have long-lasting impacts on the brain. This is harder to demonstrate in humans, but scientists believe it may also affect our own mental health in the same way. *M. vaccae* has also been shown to boost the immune system and protect against cognitive dysfunction.

M. vaccae is not alone – the tens of billions of microbes which inhabit our bodies have profound impacts on our health. It is estimated that we have at least as many microbial cells in our body as human cells in our body, and they affect everything from our immune system to our behaviour. Our gut microbiome, for example, has been linked to inflammatory bowel disease, asthma, autoimmune disorders and autism spectrum disorders. These conditions are on the increase, something scientists believe is connected to our lack of exposure to 'old friend' microorganisms in our modern, sterile world. Microbes have co-evolved with the humans for millennia, and analysis of DNA fragments in fossilised human faeces shows a much greater diversity of bacteria than are found in our guts today, including many unknown species. This suggests that the human gut has experienced an 'extinction event' in the last millennium, potentially damaging our mental and physical health.

Other animal species are facing the same problems, particularly those involved in captive-breeding programmes.

As well as being treated with insecticides to remove parasites, animals may be given antibiotics, killing beneficial bacteria alongside harmful ones. Captive animals tend to be fed a less varied and more processed diet, and are kept in more hygienic conditions. Their symbiotic microbes therefore face exactly the same challenges as ours do. A breeding programme's chance of success can be enhanced with simple solutions such as providing access to natural environments where they will encounter beneficial microbes. As our knowledge increases, interventions could be more dramatic. In a recent, innovative example, researchers used a faecal transplant pellet to help captive koalas switch their diets between different tree species.

Human and animal health is also indirectly affected by the microbes that keep our ecosystems functioning and our food growing. As well as bacteria, the single-celled organisms we rely on include algae, protists and archaea, and modern techniques have allowed us insights into their diversity. Archaea, for example, look superficially like bacteria, and weren't identified until 1977. Molecular analysis has revealed how different they are to bacteria, yet this entire domain of life previously went unnoticed. We now know that they separated from bacteria early in evolution, and the current thinking is that archaea are the ancestors of all animals, plants and fungi. Amazingly, these single-celled organisms are more closely related to us than to bacteria.

DNA sequencing has allowed tentative estimates as to how many microorganism species there could be. An influential study from Indiana University predicted that Earth could contain up to a trillion species of microbe. If this is true, we have only identified 0.001 per cent of them. Based on species count alone, the microbes in our own bodies would be a cost-effective target for conservation. The palm of your hand may hold over 150 species of bacteria, which is more than the number of mammal species in the UK. This is dwarfed by the number in your gut, which can be home to 2kg of microbes from around 1,000 species. Reducing antibiotic use and choosing a high fibre diet to please your microbes is perhaps the world's simplest conservation scheme.

This diversity of microorganisms allows the ecological processes we witness in nature to play out in miniature. Photosynthesis, parasitism, competition, predation – all the interactions we marvel at in wildlife documentaries are happening unnoticed under our feet, in the oceans and in our bodies. This activity is vital. Microbes are decomposers, releasing nutrients from decaying matter so that they can be absorbed by other life forms. They compensate for some of the damage humans cause, such as breaking down toxins and pollutants to render them harmless. Microbes also remove CO_2 from the atmosphere, and play a vital role in the carbon cycle. Microbes associated with soils and plants ultimately provide almost all human sustenance by supporting crop production and pastures. In the oceans, microbes are the basis of the marine food web. Single-celled phytoplankton capture the Sun's energy through photosynthesis, and provide food for everything from microscopic zooplankton to 30-metre whales. They may be invisible to the naked eye, but phytoplankton are so abundant that they produce approximately the same amount of oxygen as all land plants combined.

Microbes are therefore vital for our survival, and for the species and ecosystems we are trying to protect. However, they are threatened by pollution, antibiotic use and many other human activities. In 2019, Ricardo Cavicchioli and a group of thirty microbiologists published a warning in *Nature Reviews Microbiology* about microorganisms and climate change, putting us 'on notice that the microscopic majority can no longer be the unseen elephant in the room'. The vital roles of microbes reveal yet another reason to expand our horizons beyond the species that intrinsically attract us.

We would no doubt think differently about microorganisms if our eyes could see the dramas of the microbe world that play out all around us. However, we have a tendency to ignore what we can't see, and the same is true of species that are hidden to the human eye, not because of their size but because of their location. Marine sediments, for example, cover a larger area than all other habitats on Earth, yet the species they house receive a fraction of the attention we give to those we see around us. In conservation terms, this is a major oversight, as the seabed is alive with a tremendous diversity of organisms. These are collectively known as the 'benthos', and familiar examples include sea anemones, corals, starfish, sea urchins and crabs, along with weird and wonderful creatures such as nudibranch sea slugs with psychedelic patterns. Many are slow-growing and long-lived – some sea urchins, for example, can live for 100 years. Even the harsh environments of the cold, dark waters of the deep sea are home to species adapted to survive at high pressure.

Some seabed organisms have a particularly profound ecological impact – the so-called 'ecosystem engineers'. On land, we are familiar with organisms that play important roles in creating habitats, such as beavers building dams. But few of us have heard of equivalent species on the seabed, let alone appreciated their function. Mud shrimps, for example, are a group of species that affect ocean habitats around the world. They are long and thin, often with translucent exo-skeletons that give rise to their alternative name of 'ghost shrimps'. Their impact comes largely from their burrows, which vary from simple U-shapes to complex branching tunnels. Some species even create chambers and attach plant material to the walls. They then eat the organisms that feed on the decaying plants (they have been referred to as 'gardeners'). Their burrows can be up to three metres deep and ensure that nutrients and oxygenated water reach lower into the sediment of the seabed. Nutrient transportation by mud shrimps benefits other organisms, from microbes to seagrass, with effects that can ripple up to fish and birds.

The sand and mud at the bottom of the ocean also has an important benefit in carbon storage. When dead whales, sharks and fish fall to the sea floor, the carbon in their bodies can be stored in deep marine sediments. We tend to think of trees as the best organisms for capturing carbon, but seagrass, for example, captures carbon up to thirty-five times faster than tropical rainforests. And imagine how much carbon is stored in a 173-tonne blue whale. Not only will this carbon remain out of the atmosphere for the whale's life, but when its carcass descends to the bottom of the sea, that stored carbon is taken out of the atmosphere for hundreds and perhaps thousands of years. Organisms on the seabed play important roles in consuming, decomposing and burying organic matter that sinks there, but this whole process is now under threat.

Overfishing has reduced the amount of carbon stored in both living and dead fish, and has been catastrophic for other organisms too. Bottom trawling has been particularly damaging, with plants, animals and their habitats being destroyed by heavy nets designed to dig into the sediment. Along with the target species such as cod, shrimp and squid, vast numbers of non-target organisms are caught. Over half the catch can be thrown back into the sea, creating pockets of decaying creatures.

Worryingly, the seabed is also being explored by mining companies. Of particular concern is the attention given to deposits of metals such as copper, iron, lead and gold that occur around hydrothermal vents. From tube worms to squat lobsters, a spectacular diversity of species thrive around these vents, and their communities are unique in gaining their energy not from the Sun but from chemicals released by the sea floor. Life around hydrothermal vents was only discovered in 1977, and since then species have been described at an average of one every ten days. Mining could mean many species are driven to extinction before we even know they exist.

As so often is the case, the people who cause the damage won't be the ones who pay the price. Richer nations enjoy the short-term gains of extracting marine resources, with taxpayers even subsidising the fisheries that are depleting fish stocks and damaging marine habitats. Distant systems are linked, though, and our fates are entwined with ocean organisms we have never heard of. Although the sea squirts, worms and mud shrimps of the ocean floor aren't the beauties that attract conservation funding, 'out of sight out of mind' isn't an attitude we can afford. If we are seriously going to address climate change, we need to understand, respect and protect life in the ocean.

Just as deep-sea sediments are home to organisms with exquisite adaptations, so too are the world's harshest soils. Perhaps the most extreme are those in Antarctica, though animals living in the ground aren't what spring to mind when you think about Antarctic wildlife. Penguins, whales and other charismatic species attract conservation attention, and for good reason. Threats such as illegal fishing, climate change and pollution have led to some alarming declines. However, we tend to forget that Antarctica's land and sea are also home to unique miniature creatures, many of them invisible to the naked eye.

Antarctica offers some of our planet's most inhospitable environments, and until recently we didn't imagine the harshest of these could sustain life. The explorer Robert Falcon Scott, who perished in 1912 when returning to his ship after reaching the South Pole, described Antarctica's dry valleys as 'the valley of the dead'. As recently as the 1980s, it was believed that the soils of the dry valleys are sterile. This was perhaps a naturral conclusion, as the McMurdo Dry Valleys have average air temperatures ranging from −16 to −21°C and annual precipitation of less than 10 centimetres, but it turns out to be far from the truth. The micro-animals found there include tardigrades (known as water bears because of their fat bodies and stubby legs) and nematode worms.

Like hairworms, nematodes aren't closely related to the segmented earthworms we are familiar with. Instead, they are thin and smooth, and their long straight gut means that their body plan can be described simply as 'a tube within a tube'. Nematodes vary hugely in size, and parasitic species in particular can be very large. A nematode parasite of sperm whale placentas can grow to 8 metres long. More often, they measure less than a millimetre. The nematodes found in the Antarctic dry valleys are microscopic, so it is no surprise that they were long overlooked. However, research has now revealed that the soils are alive with nematodes, with one dominant species: *Scottnema lindsayae*. This nematode may be little known to the world, but it has captured the imagination of those people lucky enough to study it. Professor

Diana Wall from Colorado State University spends months each year studying the invisible wildlife of Scott's valley of the dead, and *S. lindsayae* is one of the species that keeps her returning. As she says, 'It is truly spectacular to think that this worm, looking so delicate, lives in one of the most extreme soils in the world – Mars on Earth.'

S. lindsayae was first described by Father Richard Timm, a Catholic priest whose varied career ranged from Antarctic explorations to disaster relief. The name honours two figures in Antarctic history. *Scottnema* refers to Robert Scott, and *lindsayae* to the woman who collected the first specimen. This was Kay Lindsay, a pioneer of women scientists in the Antarctic. The US Navy established an Antarctic outpost in 1959, but at the time it banned women from Antarctic travel. That changed in 1969, when the Navy lifted a ban on women travelling to Antarctica. This allowed Lindsay to begin her research there, and she was one of the six women who went by plane to the South Pole on 12 November that year. When the plane's cargo ramp was lowered, all six linked arms and stepped onto the ice together, becoming simultaneously the first women at the South Pole.

The worms are as remarkable as the people who study them. Their ingenious adaptations for survival on Earth's coldest and driest continent include the ability to enter a state of suspended animation. The shrivelled worms may appear dead, but can revive months, years or even decades later. With the Antarctic rapidly warming, it's inevitable that a species so exquisitely adapted to the cold is in trouble. Sudden warming events this century have caused surges of meltwater to seep into the dry soils which *S. lindsayae* thrives in. As a result, its population in the dry valleys declined by 65 per cent between 1993 and 2005. Other nematodes that are adapted to warmer, wetter soils have moved in and outcompeted *S. lindsayae*, but they haven't compensated for its loss, so nematode numbers have declined overall.

With so few species found in the dry valleys, the declines are likely to have a large effect on the soil ecosystem. Microorganism communities in particular will be affected, because microbial growth is stimulated by nematodes feeding on them. Nematodes also play an important role in the carbon cycle, which is particularly vulnerable to disruption in Antarctica. In other ecosystems, declines in one organism may be counterbalanced by an increase in others. However, the tiny number of species in Antarctica means there is no backup plan – we can't assume that other species will perform the same functions as 'the lion of the Antarctic'.

These nematodes are important and declining, yet they don't feature on the conservation agenda. No conservation charity will start a campaign to support a microscopic worm. And this story isn't unique to Antarctica – around the world, the overlooked multitude of nematode worms are playing vital roles. From the ocean bed to the human gut, nematodes are everywhere. Four out of every five animals on land are nematode worms, and they represent 90 per cent of all animals on the ocean floor. A recent study estimates that, in the soil alone, there are 57 billion nematodes for every person, and their combined mass is over 80 per cent of the total weight of the human population. That is a lot of carbon stored in the bodies of these tiny animals.

As you might expect for such a diverse group of species, nematodes play a multitude of roles. As consumers of microorganisms and prey for other animals, they are a key part of the food web. Parasitic species can regulate insect populations, including agricultural pests, and herbivores control the growth of microorganisms. Soil nematodes are fundamental to the processes of life, from allowing plants to grow, to decomposing them. A few species even act as 'hygiene police', controlling fungi and bacteria in insect nests.

Of course, things aren't always rosy – some nematodes cause great damage. The golden nematode, for example, threatens potato and tomato production by sucking nutrients out of plant roots. Other species cause human disease, and about half the world's population have them in their guts. Nematodes can cause serious problems, especially in children. Some conservation priorities are fairly clear. I don't think any of us are going to argue in favour of conservation of *Plasmodium falciparum,* the parasite that causes malaria in humans. If we ever can, we may well drive *P. falciparum* to extinction, just as we have for the viral diseases smallpox and rinderpest.

Conservation faces a multitude of value judgements, as we will explore over the course of this book. We need to consider every organism on a case-by-case basis. And, to make things even harder, we will never fully understand the role each species plays in every situation. When we allocate tickets for the ark of the sixth extinction, we can't predict the full implications. We can, however, use the evidence we have to make wiser choices. We can make informed decisions based on the astounding amount of information available, while still being humble enough to accept that we can't predict all the outcomes. And we have to be wary of sentiment and intuition. Our dislike of wasps – or parasites or nematodes – shows how easily we can overlook fascinating and invaluable species, and we can't afford to do that in a time of ecological crisis.

CHAPTER 3

NEW ARRIVALS

Nile tilapia vs yabbies (and tales of tortoises, parakeets and other 'invasives')

'Tilapia? I avoid them whenever I can!' Marine biologist Sara Busilacchi doesn't hesitate when I ask if she likes to eat tilapia. In this, she is at one in this with the communities she works with in the South Fly region of Papua New Guinea; the introduced tilapia fish just don't taste as good as the native species they replace – their muddy taste has few enthusiasts. When it comes to conservation policies, however, the answer is much harder. Is it good policy to allow tilapia to spread in Papua New Guinea at the expense of native species such as yabbies – freshwater crayfish that used to be common in local markets? It seems that the spread of non-native fish is one of the reasons for yabby's decline, but tilapia are encouraged in Papua New Guinea as a vital source of protein. By contrast, just across the sea in Australia, they are considered a pest and controlled by tight laws on invasive species.

Sara has seen both sides of the argument in Papua New Guinea, where she has worked for a decade. She explains to me that three tilapia species have been introduced to the country's rivers: the redbreast tilapia, the Mozambique tilapia and the Nile tilapia. The latter is the most recent introduction and the most common worldwide, with a native range stretching from Senegal to Israel. Now they have spread to every tropical country, and beyond. Their grey scales are often tinged with red, and they have a long fin running along their back, which becomes strangely reminiscent of a punk hairstyle when it is held erect. They can grow to 60 cm long, but although Sara has seen some 'monsters' during her travels in Papua New Guinea most tilapia sold in the markets are about the size of a paperback book.

Nile tilapia were introduced to Papua New Guinea in the 1990s for fish farming, and tilapia ponds can now be found all across the country. The ponds are simply holes in the ground: steep-sided pits with muddy puddles at the bottom. They are no larger than my garden pond yet can be packed with tilapia feeding on algae. The 'farms' cost the villagers nothing and Sara can see why they are so popular: 'Nile tilapia is a very strong, sturdy species, and that makes it perfect for small-scale aquaculture which requires low skills. The conditions don't have to be perfect.'

Tilapia can therefore provide vital protein even to people without any training. At the same time, the crowded ponds and lack of aquaculture skills made it easy for the fish to escape and end up in lakes and streams. They were also set free by extreme flooding in 2012. Once released, their biology means that numbers grow rapidly. A female can produce thousands of young each time she spawns, and will incubate the eggs in her mouth. Even after they hatch, the fry will keep returning to the mother's mouth for protection. Their rapid reproduction means that the population increase has been dramatic. As Sara says, 'the rivers everywhere in Papua New Guinea are full of tilapia. Wherever you go, you can catch them.'

One problem comes when tilapia compete with native species, as has happened in many other countries. In Kenya's Lake Victoria, their introduction led to several native fish species disappearing. But is this a problem? Many conservationists automatically object to their presence – tilapia were introduced by people, so they simply don't belong. But, as discussed, it is arbitrary to take a snapshot of the past and say 'these distributions are the correct ones' and the fact that tilapia were spread by people is irrelevant. Every species has an impact on the environment, and we are no different. Instead, we need to consider their impact. What benefits do they bring, and what are the problems? Does the benefit of extra protein come alongside negative effects?

This could be an issue in Papua New Guinea. Yabbies are freshwater crayfish that resemble small blue lobsters, and they are becoming increasingly rare in the rivers which Sara visits. We know remarkably little about the scale of their declines or its causes. New arrivals such as tilapia are almost certainly posing part of the problem, but there are no studies on how severe their impact is. This is a common problem – studies often show that one species declines as another spreads, but that doesn't show cause and effect; at a time of environmental change, the species may well have been declining anyway.

Just because we lack data doesn't mean we can ignore the problem. There are plenty of reasons why we might be worried about yabby declines. They formed the basis of small artisanal fisheries, bringing food and livelihoods, and were also important in the diets of native fish, playing key roles in food webs. Elsewhere in Papua New Guinea there are yabby species threatened with extinction, such as the zebra crayfish. Found only in Lake Kutubu in the remote central mountains, these are suffering from the impacts of tilapia, overfishing and pollution. It's a common story – problems caused by both human activities and introduced species – and conservation would need to address at least one of these issues to reverse wildlife declines. However, there seems to be little appetite in Papua New Guinea for curbing the spread of tilapia.

Australia takes a very different approach. Significant resources are dedicated to keeping the Nile tilapia and other species out – and for good reason. The country's unique wildlife means that non-native species have had a much larger impact than elsewhere and once they are established there is often no going back. Australia has already had experiences with tilapia. Mozambique tilapia were illegally introduced in the 1970s and have since multiplied at terrifying rates. The story goes that eight fish were imported for an ornamental pond, then more than 12 tonnes of fish were removed just eighteen months later.

As well as being fast breeders, Tilapia males are aggressive to other fish, keeping them away from nesting areas, and aquatic plants also suffer from grazing. The fish can increase bank erosion, and there is also the potential for them to spread disease and compete with native fish for food. All this combines to justify Australia's spending on biosecurity.

Papua New Guinea, however, focuses on the benefits which tilapia bring. Lack of resources means that Sara can't perform extensive ecological or economic studies on their impact, and instead relies on the knowledge of local people, through interviews and focus groups. One thing her work has revealed is that tilapia are important for women's livelihoods.

Papuan society has distinct gender roles, not least in fishing. While men hunt for dugongs or go diving for sea cucumbers, women harvest lower-value species such as tilapia. People may find yabbies much tastier, but they have accepted the tilapia because they are easy to catch. If it's a choice between tilapia and hunger, there is an obvious answer.

This realisation changed Sara's perspective: 'If you work in these communities, you realise that it's a matter of survival. They cannot do things any other way.' Communities rely completely on natural resources and local people see things very differently to conservationists. Sara says that she used to say to people, 'If you keep doing this, in ten years your son or your daughter won't be able to catch this fish.' But, even when the species had great economic and cultural value, their answer would be 'And?'. That was a problem to be addressed in the future. It's a perspective that Sara came to understand: 'For me it was inconceivable that soon this species wouldn't be around, yet the people who live with it every day accepted that it would be gone. For them, there is no other choice — they have to feed their kids tonight and pay for their school fees tomorrow.'

Alongside declines caused by overfishing, there are larger forces at work in Papua New Guinea. In 1984, an area of mountains was carved up to create the Ok Tedi mine. This open pit resembles a vast Roman amphitheatre, with a road of hairpin bends allowing mining vehicles to descend to its depths. Papua New Guinea relies heavily on mining and oil extraction and about one fifth of the government's income comes from this single mine. The environmental impact, however, has been extreme. The mine is in the headwaters of the Fly River, and mine waste discharges into the water. Sediment flows are made worse by soil being washed into the watercourse when the forest is logged, and the overall result has been a decline in fish populations. A build-up of mine tailings has also caused flooding, covering fertile land and forest with mine waste.

These impacts have been devastating for the livelihoods of local communities, something which compensation money has failed to reverse. This is a sobering thought for us all – the mine provides gold and copper, both of which are found in the electronic gadgets and cars that we buy (the average car contains more than a kilometre of copper wire).

Around the world there are countless cases of local species declining due to a combination of introduced species, over-exploitation and environmental pollution. The interplay between these factors makes it hard to know the effect of each introduced species, and their impact varies massively. There's no shortcut to knowing which species will cause problems, and many species bring both problems and benefits. One example is the zebra mussel, a shellfish so called because of its brown and beige stripes. In America's Great Lakes, they have outcompeted native species and damaged boats and infrastructure. In some ways, however, they benefit the ecosystem. They filter out pollutants and are food for species such as the round goby. This bottom-dwelling fish is another introduction, and now forms a major part of the diet of native predators.

As we make judgements about when to embrace native fish species and when to fight them, we are faced with mul-tiple questions. Do we value species more if they have been in the area for a long time? How do we weigh up good fish-ing with flourishing of native species? Is the debate different if non-native fish provide food for people with few choices? What price are we willing to pay to keep non-native species out of an area, whether that is the price of biosecurity or the loss of an accessible food source like tilapia?

These questions feed into the different judgements on the dilemma of Nile tilapia versus yabbies. In Papua New Guinea, people have favoured the tilapia, spreading them at the expense of native species. However, in Australia, yabbies are favoured and Nile tilapia are kept out. How should conservationists react to tilapia in Papua New Guinea? Should they to try and reverse its spread? And is Australia using resources wisely with its strict biosafety controls?

On the face of it, the judgements about tilapia should be the same in two countries which are only divided by a few kilometres of water. However, there is a good reason why attitudes differ. In Papua New Guinea the human population density is much higher, and in the south there has been an influx of people displaced by the Ok Tedi Mine. Most of them have no land, so rely on fish for food. Given that native fish are declining, and would be even if non-native fish were eradicated, tilapia are much-needed protein. In contrast, fishing is just one option for people living on Australian islands, and they can fish commercially for higher-value species.

Another difference is that Australia has the money to spend on biosecurity, which is a burden that Papua New Guinea can ill afford. Australia's spending on biosecurity may be huge, but it is nothing compared to the resources it would take to eradicate tilapia and other non-native fish in Papua New Guinea. To Sara, this is an important point: 'In Papua New Guinea tilapia is already there. The damage has been done because it is already in the environment. It's an easy species to farm, so it is good for the livelihood of communities, and they will continue to target it for farming.'

Even if we could eradicate tilapia completely, we are left with some uncomfortable truths about the causes of native fish declines. The challenges which local people face when trying to feed their families, as well as our demand for copper and gold, are integral parts of the problem. Eradicating non-native species would not only be incredibly expensive, but it would do nothing to tackle these challenges. More effective solutions may include community management of fisheries, for example, and support for alternative livelihoods.

The case of tilapia versus yabbies highlights the challenge of making decisions about new arrivals based on their impact; this may be positive or negative depending on who you ask, and it is surrounded by uncertainty. Ultimately, it seems clear that tilapia are going to triumph in Papua New Guinea. If the fate of yabbies is going to improve, it will be because of other conservation interventions.

Elsewhere in the world, native crayfish species are getting far more attention. Europe's crayfish in particular are receiving greater protection from the threats of introduced species. In Britain, for example, there is a complex challenge that highlights the blurred line between native and non-native. The story begins with an invader which has been unanimously judged as unwelcome: the crayfish plague.

Crayfish plague is a water mould, a microorganism which spells certain death for any European crayfish it infects. It first hit European waterways in Italy in 1860, leading to a steep decline in the native crayfish population, and was next found in Sweden in 1907. Crayfish are a delicacy in Sweden, eaten at summer parties, so their decline had cultural significance. People therefore searched for a solution, and fifty years later they believed they had found it: American signal crayfish are resistant to the plague, so seemed perfect to repopulate Sweden's waters.

What nobody realised is that American signal crayfish carry the disease, so they spread the very problem they were meant to address. The species has since spread to more than twenty European countries and continues to colonise new waterways, transmitting the plague to any crayfish who live there. They also reproduce more quickly than native crayfish, tolerate a wider range of conditions and often compete with them for food. As a result, Europe's native crayfish have struggled, while the American signal crayfish has thrived.

In Britain, the dramatic fall in numbers of the native white-clawed crayfish means it has become a focus for conservation. This is accompanied by a classic 'save our native species from the imposters' narrative, although behind the scenes white-clawed crayfish have been the subject of a rather peculiar debate. The International Union for the Conservation of Nature (IUCN) lists white-clawed crayfish as native to the UK, but nobody knows when or how they spread here.

Genetically, English crayfish are indistinguishable from those in northern France, and it is possible that humans brought them north at the end of the last ice age, while the two countries were still connected. There's no evidence confirming this, though, and crayfish may well have been a much more recent arrival. Traditionally, the church calendar contained many periods where rich foods such as meat were avoided, so monks and nuns were always on the lookout for alternatives. Beavers and seals, for example, were declared to be fish because they spend so much time in water. But a particular favourite was crayfish, and the wide distribution of white-clawed crayfish is no doubt the result of monks spreading them to new streams. Monks may also have been the ones to first transport them from France to England.

Given our inclination for a binary classification of 'native' and 'non-native', the ambiguity poses a problem. Is the white-clawed crayfish native to Britain? It depends partly on your definition, and there are plenty of options to choose from. To many people, a species is non-native if it has been introduced by humans, while to others the definition is based on when it arrived. The cut-off for arrival can be seen as ancient or modern, with common options ranging from the end of the last ice age (about 11,000 years ago) to the sixteenth century. This latter date roughly corresponds with Columbus's arrival in America, and perhaps reflects broader American attitudes (like Theodore's Roosevelt's views on wilderness being the landscape before Europeans arrived). Although 1500 didn't mark such a sudden change in European habitats, it does correspond with a time when the movement of people, and therefore species, becomes much more common. This date has therefore become a common cut-off, adopted by organisations such as the UK's Joint Nature Conservation Committee (the public body that advises the government on nature conservation).

Using a definition based on presence in the year 1500 puts us in confusing territory for the white-clawed crayfish though. The first reliable record we have is Samuel Pepys recording the 'crayfishes' he saw during a visit to Hungerford in 1668. This is too late to place them in the 'native' category. But there are earlier recipes, dating from the 1400s, that mention what seem to be crayfish.

So where does this leave us? Does it matter that if I visit Hungerford today I see crayfish with a bluish tinge to their claws, unlike their white-clawed cousins Pepys noted 350 years earlier? The cut-off of 1500 is an arbitrary baseline that won't reveal whether or not the white-clawed crayfish should be a priority for conservation funds. And the role of church fasting rules in prompting its spread puts the 'invader from America' outlook in context, and reminds us to make judgements simply based on the signal crayfish's impact.

The arrival of signal crayfish to Europe has actually brought some benefits. They fulfilled one of the original objectives of their introduction: more crayfish for fishing. And it's not just humans who eat signal crayfish – so do herons and fish. In Western Europe, introduced crayfish have become an important food source for white storks, which are currently being introduced to southern Britain. We will see whether these opportunistic eaters take a liking to signal crayfish. However, there is ample evidence of non-native crayfish causing problems for people and other wildlife: their burrows can destabilise riverbanks and displace water voles. Studies reveal a decrease in the diversity of aquatic invertebrates in the presence of signal crayfish, along with examples of declines in the number of plant species.

Attempts to eradicate introduced crayfish populations have not been promising, and even the most intensive trapping programmes have failed. Preventing their spread is much more realistic, and where they already exist we can try and reduce their impact and preserve our white-clawed 'native'. The solutions to help white-clawed crayfish can be relatively simple, such as providing bricks with holes for them to shelter in, or more extensive habitat management along the banks. Other interventions are nothing to do with introduced species: water pollution is a threat, for example, and tackling this would have wide benefits for many species. White-clawed crayfish have also been welcomed in Scotland, where they were only introduced in the mid-nineteenth century. They may not be native, but the places where they occur have been termed 'ark sites', to show their importance for crayfish survival.

White-clawed crayfish are by no means the only species introduced intentionally to new ranges for conservation.

In America, for example, the seeds of a critically endangered conifer from Florida are being spread as far north as Vermont to save it from fungal pathogens. In the UK, scientists moved two butterfly species north in order to show that translocations could be a cheap and effective conservation tool. As climate change pushes many species closer to extinction, some might only be saved if they are transported to new areas. These species can bring wider benefits in their new territory, as shown by giant tortoises in the Seychelles.

The Indian Ocean nation of Mauritius is home to a beautiful diversity of species, from red and green geckos to glowing reef fish. Many of its unique creatures and plants are facing extinction, and losses so far include the dodo, an owl, a giant skink and two species of giant tortoise. The tortoises had no natural predators, so when European colonists arrived they found docile animals present in huge numbers, which made for nutritious meals. Like the dodo, the tortoises' fates were then sealed by destruction of their habitat. During centuries of occupation by the Dutch, French, then British, forest was replaced by sugar cane plantations, and a whole range of species were introduced. Ecosystems were changed beyond recognition, and many species survived only on smaller islands.

When conservation efforts began in the late twentieth century, non-native species presented a big challenge. Islands have been the hardest hit by introduced species – seabirds often nest on the ground, for example, which is fine when there are no predators, but a problem when rats eat their eggs. However, the problem wasn't simply the species that the Mauritian islands had gained; it was also the species they'd lost. Giant tortoises had been important herbivores, and the island vegetation had changed because of their loss. So what would happen if giant tortoises from the Seychelles were released in Mauritius? The first scientists to suggest this were seen as mad, but their plans were a far cry from casual species releases which had happened in the past. Careful experiments were performed to discover the tortoises' impact, and there was the get-out clause that tortoises are easy to catch if things go wrong.

The introductions went ahead, and tortoises from 1,600 kilometres north now roam freely on two small islands off the coast of Mauritius. Their impact has been impressive: they are reducing the amount of non-native vegetation and are spreading the seeds of native trees such as ebony. Forest is regenerating, and the tortoises' love of introduced plants has been a cause for celebration. Of course, it isn't all good news. Non-native species such as passion flowers are an important food for Mauritian pink pigeons, which narrowly escaped extinction when their population dipped to just ten in 1991. Passion flowers are favourites of the tortoise, so grazing tortoises threaten the pigeons' food supply. Conservationists hope to solve this problem by helping the pink pigeon transition to a diet of native plants.

The impacts of the tortoise, both positive and negative, seem very clear. They do, however, raise the question of what the project is trying to achieve. Given that the project revolves around introduced tortoises, it would be ironic if its sole objective was to reduce non-native plants on the islands. However, there may be many advantages of ecosystems containing fewer non-native species, even though there's no intrinsic reason why native species are superior. For example, rare species may make a comeback, and their important functions along with them. The Mauritian islands which tortoises now inhabit are refuges for endangered reptiles, and it's possible that extinctions could be avoided due to the habitat modifications made by tortoises. And, in a world where the same species are repeatedly becoming established in new places, we may simply prefer seeing diverse habitats with different species and not a homogenised planet. Ultimately, this project shows that the careful release of species into new areas may help us achieve the future natural world we want to see, even if we've not yet agreed what this ideal future looks like.

One thing we can be sure of is that the future won't look like the past. Nobody knows what the Mauritius landscape looked like before the arrival of non-native species – black rats reached the island from shipwrecks before the first recorded landing. The flow of new species picked up pace when the Dutch arrived, with the French and British bringing yet more species as food and as stowaways.

It's a common story – colonists have lots of introduced species to answer for. When European settlers followed Columbus to the 'New World', they brought with them some of nature's most destructive organisms. Within 150 years of Columbus's arrival, up to 95 per cent of the Native Americans had been killed by diseases such as smallpox, measles, flu and typhus. This changed the course of civilisation, and the new trajectory shaped the continent's wildlife. However, not all the life forms which accompanied the first intrepid Europeans on their transatlantic voyages were quite so destructive. Among the organisms welcomed by the Native Americans was greater plantain. It is a common sight on lawns and verges, and is distinguished from other plantains by its oval-shaped leaves and green flower spike. Greater plantain is native to Europe and parts of Asia, and its spread around the world has been spectacular. As soon as it crossed the Atlantic, Native Americans added plantain to their extensive range of herbal remedies, using it to treat conditions as varied as skin irritation, toothache and intestinal troubles.

It's possible that the initial introduction was intentional – plantain's medicinal properties were well known in Europe. It was described by the Greek physician Dioscorides in the first century AD, and it later appeared in Shakespeare's plays: Romeo recommends plantain leaf to Benvolio as treatment for his broken shin. Whether or not plantain was brought to America because of its medical uses, once it was there the spread was unstoppable. Each plant produces 20,000 tiny seeds, and these are transported in hay as well as on clothes, muddy shoes and the fur of domestic animals. Plantain's colonial progression was so marked that the Native Americans named it 'white-man's footprint'.

Modern studies have supported claims of plantain's medical properties; it can indeed relieve pain and act as an anti-inflammatory. But our main use of plantain is as a vitamin-rich foodstuff, for both people and livestock, and for many wild species. Taylor's checkerspot, for example, is an American butterfly that has come to rely on ribwort plantain, a narrow-leaved species which had been introduced from Europe. This elegant butterfly, whose wings are decorated with checks of black, orange and white, had lost its habitat as prairies were replaced by farms and cities, to the point where it faced extinction. In the 1980s, it suffered a further blow as the summers became longer and drier, causing native plants to shrivel up. Ribwort plantain, on the other hand, remains green during drought, and provides an ideal Taylor's checkerspot food plant.

British scientists Chris Thomas and Mike Singer studied California's butterflies in that decade as they switched from laying eggs on native plants to plantain. The pair documented a remarkable moment of rapid evolution, which may have made the difference between survival and extinction. Chris has since become a vocal advocate of keeping an open mind about introduced species, something he explores in his book *Inheritors of the Earth*. It is important to welcome a species when it brings benefits to a new region. As ever, though, plantain spread isn't always positive. The species competes with grass on the neat turf of lawns and sports pitches and is often treated with herbicide.

The environment in towns and villages is a stark contrast to the forests and fields they replace, and species need different adaptations to survive there. Native species can't always withstand the harsh conditions of urban areas, so welcoming non-native species can increase the wildlife around us. Scandinavian countries, for example, are home to limited options for native trees that can grow in paved environments, whereas introduced trees can thrive. Urban trees bring huge benefits to people and other wildlife, so non-native species can fill an important niche.

Cities can often be an ideal place for introduced species to flourish, though not all of them are popular. In 2011, residents of London's Isle of Dogs received unexpected visits from government officials. These visitors were prepared with guns and cherry pickers to rid the UK of its monk parakeets, and needed access to gardens where this South American parrot was living. The area had been home to a small population since the early 1990s, as had the nearby town of Borehamwood, and initially they attracted little interest from authorities. That changed in 2007, when a government agency concocted a plan to eradicate the hundred or so birds living wild in the UK. Trials began discreetly a year later, and revealed that shooting the parakeets was much more effective than trapping them. By 2011, plans for the eradication had been approved by government ministers. The story had been developing behind closed doors, with no public announcement, so the first people heard of it was when the authorities asked permission for access in order to shoot birds and destroy nests.

Unsurprisingly, not everybody was happy about this. Monk parakeets are known for their beauty and intelligence, which makes them popular as pets and appealing as garden visitors if they escape. They are a striking bird, their bright green plumage set off by a dusty pink beak and a flash of teal in their wings. The government, however, feared the consequences if parakeet populations were allowed to grow unchecked. Monk parakeets are regarded as an agricultural pest in their native South American range, causing more than £1 billion pounds of damage each year. This led to concerns that they could spread to rural areas in the UK, and cause destruction in orchards and fields.

Such problems don't seem to have arisen, so far, in Europe or North America. However, in the US, the parakeets have had an unexpected impact – on electrical infrastructure. Monk parakeets are unique among parrots in building nests rather than occupying existing cavities, and they do it in style. Their communal nests are a chaotic mass of sticks with twenty or so chambers, and power lines are a popular place to build them. Given that they can be 1 metre wide and weigh 200 kilograms, these nests can cause transformers to short-circuit or overheat.

America's experience of disrupted electricity supplies and fire hazards means it was not surprising the UK government was cautious about parakeets. The protesters, however, countered the argument by pointing to the differences in electrical infrastructure: in British towns the infrastructure is largely below ground, whereas the 110-volt system in the US, needs thicker cables, and many transformers are also above ground. So far, the parakeets haven't nested on telegraph poles in London (or on the Eruv poles which mark the area where Orthodox Jews can carry objects on the Sabbath).

Another disagreement was over whether the parakeet population is likely to grow and spread. Monk parakeets have had mixed success establishing populations in other parts of the world. They have boomed in parts of the USA, Mexico and southern Spain, but populations haven't grown elsewhere and many have died out of their own accord. The UK government report pointed to the populations which have grown rapidly, while their defenders drew comparisons with populations in similar climates to the UK. Monk parakeet numbers may have soared in Barcelona and Texas, but they are relatively stable in New York State.

There was also a debate about whether monk parakeets affect other wildlife. They rely on food provided by people, both from feeders and the fruit and seeds from garden plants, and these resources are also important to native birds. This prompted a government agency to argue that 'competition for food may be an issue since monk parakeets are known to dominate feeding areas and act aggressively to competitors'. However, no research has investigated competition with native species anywhere in its new range. Even where populations have grown, we're yet to see clear evidence of an impact on other birds. But how much risk is too much?

There were lessons from ring-necked parakeets, which are already established in cities around Europe, including Amsterdam, Brussels and Paris. Stories of their origin in London vary, with one popular tale being their escape from the set of the 1951 film *The African Queen*, starring Humphrey Bogart and Katharine Hepburn, another that they were released by Jimi Hendrix in the 1960s. However, their presence in many cities suggests the more mundane conclusion that pets have repeatedly escaped or been released. Initially, the population remained low, with numbers only reaching a few hundred by the early 1990s. This began to change in the mid-1990s, though, and the London population has now reached an estimated 35,000.

Given that other European cities are likewise home to thousands of parakeets, it is perhaps surprising that there's so little evidence of a negative impact on other species. Like many birds, ring-necked parakeets are aggressive if they believe that their nests are being threatened. They will have a go at little owls, and small flocks have been seen mobbing larger species such as herons and squirrels. They can also be aggressive at bird feeders and when competing for places to nest; many European species nest in tree cavities, and parakeets aren't afraid of using force to evict them. Sometimes they can secure the cavities simply because they breed earlier than many other species. When scops owls return to Europe from Africa, for example, they may find that the best crevices for breeding are already occupied.

Sightings of aggressive behaviour and competition for nest space have caused alarm, but they don't necessarily translate into population declines. If we take the scops owl, for example, studies in Italy found no evidence that competition with parakeets reduces their number. They might not have the best places to nest, but that doesn't stop the owls from breeding. Likewise, although parakeets are known to kill bats and small birds such as house sparrows, this may not reduce the populations. Numbers are often determined by other factors, such as food, and this will be a limiting factor regardless of aggression by the parakeets. For the vast majority of native wildlife, the verdict is out as to whether ring-necked parakeets affect their numbers.

There is, however, one exception: the greater noctule bat. With a wingspan of up to 46 centimetres, this skilled flyer is the only bat that can catch birds on the wing. It has a wide distribution in the Mediterranean, and reaches as far east as Kazakhstan, but it is a threatened species, listed as vulnerable on the Red List of Endangered Species. The largest known colony of greater noctules is in María Luisa Park in Seville, an area of fountains, lakes and avenues shaded by foreign trees, where the bats find their roosts. Here, they now compete for tree cavities with ring-necked parakeets, and that has caused numbers to fall sufficiently for Spanish conservationists to support a proposal to cull the parakeets. Shooting is by far the most effective way to do this, but also the most unpopular – the scientists even faced attacks from animal activists. A political battle delayed action, during which time parakeet numbers grew. Eventually a compromise was reached allowing trials of control methods such as egg removal. However, such techniques haven't worked elsewhere and seem unlikely to solve the problem.

Although arguments about parakeet culls seem to revolve around evidence of their impact, in reality the debate is about different world views and priorities: how much each side values the birds. In the case of the government, the answer is 'not at all'. They don't value the lives of individual birds (after all, nobody complains if they trap rats or mice), nor do they value the presence of the species; as non-natives, the parakeets are expendable. If the government doesn't value the birds, then they are free to look at this from an economic perspective. The cull may be costing hundreds of thousands of pounds, but if the parakeets aren't culled they could theoretically cost far more in damage. The residents, on the other hand, value the parakeets – for the pleasure their presence brings and for the welfare of individual birds. Why would you shoot wild animals on the off chance that they damage electricity pylons in the future?

In Britain, the debate turned bitter, hitting the national news when the police came to a protester's house after he photographed officials using cherry pickers to disturb nests. Neither side changed their views about the cull, and the result was an uncomfortable stalemate. The parakeet population was reduced by removing birds, some of which were rehomed, but the total eradication couldn't take place. Many residents prevented parakeets from being caught or killed in their gardens, and the local council in Borehamwood restricted parakeet management on its public land.

Conservation organisations such as the Royal Society for the Protection of Birds were sitting on the sidelines for this debate. They did, however, share the government's awareness that, while the birds might not be causing a problem now, if we wait before we act there may be no way back. Eradicating a hundred birds is feasible, but if the population grows ,then it will be expensive or even impossible. Madrid recently announced plans to kill thousands of monk parakeets, and the long-term battle to keep the population down will be a strain on the city's resources. Such knowledge can prompt calls to act fast, not waiting to learn whether an introduced species will cause any problems.

A 'guilty until proven innocent' outlook is common among conservationists, partly because the most dramatic stories of guilt are so memorable. Although a high proportion of horror stories come from islands, havoc can still ensue when non-natives establish on mainlands. In the UK, for example, Himalayan balsam outcompetes native plants, and puts rivers at risk from both flooding and erosion. It arrived here in 1839 without natural enemies, meaning it could grow unchecked, and it has extensively colonised riverbanks, waste ground, ditches and damp woodland. It outcompetes native plants, and when it dies back over the winter it leaves bare riverbanks that are vulnerable to erosion. Another scare story comes from the south-west USA: lionfish are voracious eaters with poisonous spines, and in some places they now outnumber native fish.

However, most mainland stories are of introduced species having no discernible impact. Only about one in ten of the species which establish outside their native range become 'invasive'. In the UK, little owls, poppies, brown hare, snowdrops, Roman snails and fallow deer have all become accepted parts of our wildlife.

Taking action early therefore has both the potential to avert future disaster or to waste money in an unnecessary fight. One pertinent example is the saltcedar tree (also known as a tamarisk) which was introduced to the USA only to become villainised. Some scientists even referred to them as 'evil', citing concerns that they deprive other plants of water and reduce habitat quality. In particular, fears were raised that the trees could have a negative impact on the southwestern willow flycatcher. This small yellowish bird, unremarkable to the untrained eye, is an endangered subspecies of the willow flycatcher, and there are just a few hundred pairs remaining. Millions of dollars were spent in attempts to remove the trees, both mechanically and with herbicides, and a beetle was introduced to eat them.

However, the flycatchers turned out to be happy to nest in the saltcedars, and concerns were even raised that the introduced beetle deprived the flycatchers of nesting sites by eating saltcedar leaves.

The landscape was definitely changing, with saltcedars becoming common while many native trees declined. However, the varied reasons for the changes include dams, which alter flooding regimes. If you remove the saltcedar then native trees may not grow back, and having fewer trees is arguably worse than having an area full of non-native trees. This is a prime example of where a return to the past is not possible. The species which thrive have changed and the interactions between them have changed. Regardless of when a species arrived in a region, the important question is 'What will be its impact now?'

We also have the issue of how we define 'problem', and conservationists tend to have a broad definition of guilt. Sometimes the very presence of introduced species makes them guilty. The British countryside gives a sense of identity and place, and parakeets can threaten that. And, to many of us, changes to populations of native species is a sign that an introduced species is in need of management. We do need to be aware, however, that there is no 'correct' level – the population sizes of native species are determined by the human environment. There's nothing 'natural' about population sizes, particularly in the city. We are setting our baselines at a point long after trees have been replaced by houses and wild food sources have given way to bird feeders.

We could look at our reactions to other species to guide us in our attitude to parakeets, but these are often inconsistent. Take the yellow-legged gull, for example, a large and noisy seabird found in high numbers on the coastlines of southern Europe and northern Africa. It competes with other seabirds for nesting sites, and kills chicks and lizards, so in many places its numbers are controlled by culling adults or sterilising eggs. Further north in Europe, the closely-related herring gull attracts similar measures. Domestic cats, on the other hand, are allowed to roam freely despite consuming billions of songbirds and small mammals each year in the UK alone. Native gulls are therefore receiving harsher treatment than non-native cats. Is the conclusion that we are happy to control species which harm other wildlife, as long as we don't like them too much?

Perhaps the best analogy to parakeets is the feral pigeon. Like parakeets, pigeons have colonised cities around the world, and their long-term presence means we have a better understanding of their effects. Large quantities of pigeon droppings can kill vegetation, but it seems that their ecological impact is minimal compared with their effects on infrastructure: pigeons around the world damage stonework on historic monuments, which are often topped with spikes to deter pigeons from landing. Attempts to reduce actual numbers can be futile, because they breed so quickly, but this doesn't stop people trying. Pigeons are often shot or trapped, or deterred using methods such as electric shocks.

Unlike parakeets, culls of pigeons aren't often opposed. Pigeons are an accepted part of our cities, but seldom celebrated. We could see them as a story of nature thriving, a triumph of humankind's positive impact. Instead, we barely recognise them as nature. It's worth stopping to consider this – from the Australian cuttlefish to the satin beauty moth, there are many examples of species which thrive thanks to humans. Why do we so often dismiss them? One reason can be a dislike of non-native species. The militant language of 'the war on invasives' sometimes speaks more of xenophobia than of valuing nature and the great services it provides.

Around the world there are countless debates about whether a species should be welcomed in a new range. As Chinese water deer decline in their native Asian range, should they be protected in their new European strongholds? Should we try to control grey squirrels in the UK so that red squirrels can make a comeback, even though they are common elsewhere? Did Swedish scientists use funds wisely when they introduced a butterfly (Oberthür's grizzled skipper) to a range that may be more suitable in a warming climate? There are no clear answers to any of these questions – we can't determine where species 'belong' by looking at their past distribution or their method of arrival. Nor can we assume that every new arrival will cause problems. Once again, we need to consider each introduced species based on its particular impacts.

CHAPTER 4

NOAH'S ARK

Ko'ko' birds vs the Guam tail louse (and a look at seedbanks and modified chestnuts)

In Guam, you are never far away from an image of the rail, a chocolate-brown flightless bird similar in size to the common moorhen. Known locally as the ko'ko', it has become a national icon, appearing as a logo for sustainability initiatives and in promotional materials for tourists. Its rise to fame began in 1984, when a line of people assembled elbow-to-elbow on the Pacific island of Guam, with the aim of catching every remaining rail. Conservationists and volunteers swept through the dense forest, collecting all the rails they could find – adults, chicks, and even eggs. They were successful, up to a point. In total twenty-two rails were captured – and in 1987, when the programme concluded, the US Fish and Wildlife Service declared the species 'extinct in the wild'.

The move to captivity could easily have been the end for the rail, and many believed it would be. But this secretive bird of the forest floor had captured imaginations, and the conservationists did everything possible to save them. So they built cages, took measurements, set up breeding programmes and cleared the birds of their parasites. This last action is standard for birds in captive-breeding programmes, and was done without a second thought.

The Guam rail louse, unlike its parent species, is not part of Guam's PR. Indeed, nobody has shown much interest in its fate. I first came across it in a paper in *Oryx*, the Flora and Fauna International journal. This listed all louse species facing extinction along with their hosts, and stated: 'We have no information about the fate of *Rallicola (Rallicola) guami*, a louse species known only from the Guam rail.'

The authors classed the Guam rail louse as an example of 'conservation-induced extinction', because they assumed the last lice had been destroyed when conservationists deloused the captive birds. Perhaps as shocking as its extinction in the hands of conservation was the question mark next to this conclusion: not only might we have caused an extinction, but we didn't even record it. This uncertainty left me uneasy, and in an attempt to clear it up I contacted Laura Barnhart Duenas from the Guam Department of Agriculture and her colleague Suzanne Medina, who are key players in the Guam rail recovery. My question came as a surprise: 'When we got your email about the louse, Suzanne and I just looked at each other,' Laura says. 'We remembered vaguely someone mentioning it, but it wasn't really on the radar. The focus back then was saving the birds.'

If the people responsible for the Guam rail breeding programme didn't know anything about the louse, it seemed likely to be extinct. Laura didn't even know if a specimen had been preserved, and coronavirus restrictions meant that she couldn't visit Guam's insect collections to check. The louse was of so little importance that we don't even have records of a scientific specimen. So how did we come to lose the louse and save the Guam rail, and does this matter?

The story of the Guam rail and its struggle to survive starts with a single US military plane landing in Guam in 1949. The plane took off from the Solomon Islands, and hidden in its undercarriage was a pregnant brown tree snake. This one snake was enough to form a breeding population, feasting on the ready food supply that included chicks and eggs. Brown tree snakes have no predators on Guam, so their numbers grew to millions. Within a few decades, the snake eradicated eight of the island's bird species and caused the populations of others to plummet.

This turned out to be one of the worst examples of damage by an introduced species. Without the birds to perform functions such as dispersing tree seeds and controlling spiders, the forest thinned out and filled up with webs. Other introduced species brought problems, too – insects attacked the native cycad tree. With changes happening so fast on Guam, it wasn't instantly clear why the rail was in rapid decline. The island's bird populations had only recently been monitored and there was no data on the spread of the snake, so biologists relied on local knowledge to work out what had changed. For the locals, it was ever more clear, as they were having to kill snakes to protect their chickens. They recounted how the snakes had gradually spread northwards from the military base in southern Guam, and the rails simultaneously retreated.

This knowledge showed that reintroductions on Guam at that time would be doomed to failure and prompted conservationists to gather up the last few wild birds. While tree snakes were still at large, released birds would face the same threats they had been rescued from. 'People didn't want to keep these birds in captivity,' Laura says. 'They knew they had to start looking right away for a place to release the birds, or there would be a gridlock where there's no space in zoos and nowhere for the birds to go.' The team were conscious that this dead end was already being faced by many other species confined to zoos. Sometimes there are no options for release even after decades of captive breeding, and interest fades. As Laura says: 'Every year that passes by, people lose a little hope.'

The search was on for a suitable release site, and the ideal choice seemed to be Rota island, just 100 kilometres to the north of Guam. Rota has a similar habitat and is free of snakes. There was, however, a problem: as far as anyone knows, Guam rails were never found there. Rota once had its own rail species, but this was an ancient extinction so there was no recent history of a rail forming part of the ecosystem. Despite this drawback, the release project went ahead in 1989. 'Some of the regulations back then helped us,' Laura recalls. 'If someone was to do this now, there would be a lot more red tape, a lot more hurdles.'

In this instance, the gamble paid off. Rota now has a self-sustaining population of around 200 rails and populations are topped up each year, with children from local schools taking part in release ceremonies. In 2010, rails were released on a second snake-free site: Cocos, a tiny islet off the south coast of Guam. These efforts culminated in one of conservation's greatest success stories. In 2019, when the International Union for the Conservation of Nature (IUCN) released the updates to its Red List of Endangered Species, the Guam rail was reclassified from 'extinct in the wild' to 'critically endangered' – only the second bird to have been downgraded in this way.

The other bird to achieve this accolade is the California condor – and it had a very similar journey. In 1987, the same year that the final Guam rail was captured, the US Fish and Wildlife Service caught all the remaining California condors. These black vultures – North America's largest land bird, with a massive 3-metre wingspan – faced extinction due to poaching, lead poisoning and habitat destruction. Just like the Guam rail, their time confined to captivity was short, and their successful reintroduction began in 1992.

They have another similarity, which usually goes unnoticed. Like the ko'ko', the condor had its own specialist louse. And, likewise, when the condors arrived in zoos they were liberally dusted with insecticide to remove their parasites. For the California condor louse, this meant extinction.

If conservation causes the extinction of louse species when we protect birds, we need to ask two questions. What are the practical implications of the losing the louse and of saving the bird? And did the louse's intrinsic value make its loss morally wrong?

To answer the first question, it helps to take a brief look at the long history of extinctions caused by humankind. When our ancestors first arrived in Europe, they found cave lions, giant polar bears and woolly rhinoceros; in South America, there was a hippo-sized giant ground sloth, while North America was home to lions, giant tortoises and sabre-toothed cats. These megafauna are long since extinct and their loss had a series of knock-on effects, changing habitats in ways that benefited some species – humans among them – yet caused others to decline. Much as I am saddened that I will never see a *Diprotodon,* a bearlike marsupial the size of a hippo, I'm glad that I don't spend my life in fear of huge animals eating or trampling me. We are thriving despite these extinctions. However, even the impacts of ancient extinctions may come back to bite us, as Sergey and Nikita Zimov remind us with the melting of the Siberian permafrost. Given the ways in which other species support our lives, the pace of extinctions we see today should terrify us. It is estimated as being up to 1,000 times what we would expect from the fossil record. We are losing two vertebrate species each year and numerous plants, insects and fungi. Each loss brings risks of big changes.

Even losses of seemingly inconsequential species can have a great impact on the wider ecosystem – something that wasn't so well understood when the decision was made to save the Guam rail and eliminate its louse. Sometimes other species will fill a niche, so changes will be very minor, but a whole ecosystem can change from an extinction – and predicting such impacts is harder with climate change, as species that currently play a minor role may become important.

We are not, however, helpless in the face of uncertainty. We know that the loss of certain species can have a huge impact on ecosystems, and some of these we can predict. Africa's forest elephants, for example, have declined massively since European colonisation. By consuming vegetation and dispersing seeds, elephants change the structure of the forest, potentially enhancing carbon storage. Their ongoing loss means a decline in these functions, and if they were to go extinct then this would be irreversible.

There are also many marine examples. The humphead wrasse, a reef fish which gets its name from the bulge on its head, grows up to two metres long and is a predator of snails, starfish and crustaceans. It is endangered and, as numbers fall, we are losing its service of protecting coral reefs from destruction by crown-of-thorns starfish. The loss of corals in turn exposes coastlines to storms and erosion.

We know less about the importance of smaller species such as lice, as they don't attract the same kind of attention. But we saw earlier that parasites can have surprising advantages, and it is possible that the Guam rail itself would have benefited from the louse. Specialist parasites, dependent on a single host species, tend to be more benign than those that infect lots of species. Without them, the birds could be more at risk from generalists which cause more damage. Such questions never crossed anyone's mind when the Guam rails had their parasites removed. 'It speaks to the time,' Laura says. 'People focused on saving the things they could see. Unfortunately the louse was lost in the fray.'

Changing attitudes mean smaller things are now gaining more attention and their ecological roles recognised. In Guam, insects are now monitored, and the island's conservation targets include dragonflies and tree snails. Laura thinks that decisions about the rail's parasites might be different if they were made today.

Saving the Guam rail reveals another aspect to setting conservation priorities, one we shouldn't underestimate. The rail was chosen largely because of its cultural significance. On Guam every child is told the story of why the ko'ko' has stripes on its belly: a tale of how a young monitor lizard was granted her wish to sing like the birds, and finds herself singing so beautifully; then wants to be as bright as the forest birds and asks the ko'ko' to decorate her with spots, and in return offers to liven up her plumage by adding some stripes. The ko'ko' covers the lizard's body with tiny yellow spots, and the lizard starts painting stripes on the ko'ko'. However, she gets tired and stops, and, enraged, the ko'ko' bites the lizard's tongue, so she can no longer sing.

The lizard's forked tongue and croaky voice are a reminder of the perils of vanity and not being true to your word.

The connection between people and rails goes well beyond stories. Before the arrival of the brown tree snake, rails had been common on Guam, so older members of the local Chamorro population remember them well. Laura enjoys listening to their stories and tells me that older folk 'talk about how ko'ko's used to be their pets – wild birds would come into the house and sleep on the pillow with them'. This is in keeping with Laura's experience of raising rails in captivity. A chick she reared by hand went from a black ball of fluff with legs as long as its body to a feathered adult who snuggles up next to her and cleans the dust from her hands. Her colleagues, too, have anecdotes about the rails they have raised: 'One of our technicians lost his glasses in an enclosure and the next day a ko'ko' brought them up to the front for him. They love shiny things!'

The relationship between local people and the rails also has a more practical element. During the Second World War, Guam faced food shortages, so locals relied on what they could find in the forest. The ko'ko' became an important food source. Food now arrives on the island in regular shipments, but the loss of local wildlife has left people feeling vulnerable should that change. For Laura, bringing back lost wildlife to Guam has another dimension: 'After reading the accounts of all our native species, it would be really amazing to look up and see a fruit bat back again in the tree or be able to see a ko'ko' run across the road. It would enrich my life.'

Like Laura, we all want to live in a world where the most charismatic species continue to exist. However, the value that we place on a species simply because it exists shouldn't be confused with intrinsic value, which remains even if there is no one to value it. Intrinsic value is a property of the species in question, regardless of its impact on humans or any other organisms. It's an important distinction. If species have intrinsic value, then conservationists hold a trump card: the moral imperative to delay extinctions could outweigh any costs of conservation, whether financial, social or in terms of animal welfare. If this value is assigned by people, then it is legitimate to disagree about the importance of a species. People who are asked to leave their land to protect endangered species, for example, may justifiably question why a species is seen as more important than them.

Any argument based on intrinsic value should give equal weighting to all species, however small, however ugly. Given that we don't have the resources to protect each of the million species threatened with extinction, a strategy based on intrinsic value won't allow us the luxury of protecting favourites. We would instead prioritise whichever species are most effective to save.

It therefore might come as a relief (albeit an uncomfortable one) that attempts to justify the intrinsic value of species have been fraught with difficulties. As we have seen, humans are not separate from nature and there's no reason to believe that human-induced extinctions are any less natural than those caused by asteroids, volcanoes or other organisms. Each loss paves the way for something new – in fact, some scientists believe that the five previous mass extinctions have ultimately increased the diversity of life. If it wasn't for extinctions, we wouldn't be here, and who knows what future organisms will evolve as a result of any human-induced extinctions.

Being natural, however, isn't enough to make extinction acceptable – humans can't justify all crimes with the excuse that we are part of nature. If you caused a person to suffer, that is widely agreed to be wrong. Our focus on suffering makes it easy to see when we are harming another person or animal. The same isn't true of a species which has no capacity to feel pain or pleasure. It is not clear that it's possible to harm a species, beyond the harm caused to individuals. As philosopher Elliott Sober points out: 'What do species want? Do they want to remain stable in numbers, neither growing nor shrinking? Or, since most species have gone extinct, perhaps what species really want is to go extinct, and it is human meddlesomeness which frustrates this natural tendency?'

A belief that species have intrinsic value is more of a comment on how humans perceive the world than on where value truly lies. We could just as logically place our value at the level of a gene or genetic code, arguing that because a DNA molecule has evolved in a particular sequence it has intrinsic value and we are morally obliged to preserve it. But we don't see the world in that way – until recently, we didn't even know that DNA existed. Instead, we value categories of things we can easily see, as our long tradition of grouping plants and animals into species reveals. These groupings, however, can be subjective.

A common species definition focusses on the ability of individuals to produce fertile offspring, but this doesn't always make for neat categories. The mallard duck, for example, can breed with over sixty other species. One of these is the pintail, an elegant duck easily recognised by its pointed tail sticking up in the air, yet nobody is arguing that the mallard and the pintail are the same species. Definitions get even fuzzier in the realm of plants, particularly as most reproduce asexually, so a definition based on interbreeding doesn't work. For plants that do reproduce sexually, crosses between species are common. When two plant species interbreed, their offspring can inherit an entire genome from both their parents. This hybrid plant may be unable to breed with either of its parent species. Two species have therefore combined very quickly to create a new species, which is believed to be a major way for new plant species to arise. All this begs the question of how 'species' can be morally significant if definitions are arbitrary.

These neat categories also don't hold true over time. Before Darwin, the stability of species was taken for granted. However, Darwin came back from his voyage on the *Beagle* determined to explore a radical alternative to this orthodoxy. Species didn't exist now exactly as God had created them, but changed over time. When he announced this revolutionary idea to the world, his contemporaries were forced to re-evaluate their very identities. This caused almighty problems for creationists, of course. But it also poses problems for conservationists. If species don't exist as neat categories in nature, why do we place our value there?

All these arguments are nails in the coffin of singling out 'species' as a place to ascribe intrinsic value. They combine to give a clear answer to the most famous thought experiment in environmental ethics: 'the last man' (perhaps better renamed 'the last human'). This been told in many ways, and aims to clarify whether anything in nature has value beyond its impact on organisms which can suffer:

Imagine this. You are the last person alive, and you share the Earth only with plants and microbes. On your dying day you decide to entertain yourself by cutting down the last remaining oak tree, just because you can. Are you morally wrong?

An answer of 'yes' would indicate that species have intrinsic value. But, as we have seen, a logical analysis brings us to an answer of 'no'. The world simply loses a category (oak tree) that was defined and valued by humans, a loss that was at some point inevitable anyway. There has been no suffering, and no pleasure denied. For some people, the conclusion that species don't have intrinsic value is counterintuitive or uncomfortable, but it needn't be. In many ways it should be liberating – we are free to discuss logically what we should save and why, and not just fight an anti-extinction battle that is doomed to failure.

Once we are free from the idea of species having intrinsic value, we don't have to give them equal weighting in conservation. Instead, we can analyse the impact of saving the Guam rail and use this knowledge to decide whether it is wise use of conservation funds. The benefits, however, are just part of the equation – we also need to look at the costs of preventing an extinction, and the likelihood of success. Many were initially sceptical about the Guam rail's chance of survival, and Laura still encounters people who doubt it has a future. This is partly because Guam itself is still not suitable for release – the rails living on Guam are kept in enclosures to protect them from snakes, cats and other predators. However, Laura believes that one day the ko'ko' can roam free on the island. Conservationists are actively trying to reduce snake densities by dropping dead mice laced with poison from helicopters. So far this seems to be working, and, even more importantly, rails seem to be thriving in such areas, even though some snakes remain.

Another issue to consider is climate change. Because of its elevation, Guam is not quite as vulnerable as some of its neighbours in the Pacific, but all island nations are feeling nervous in the face of sea level rise. Guam and the surrounding islands have already seen changes to their rainfall patterns, and experience storms outside the usual typhoon season. For this reason, Laura and her colleagues have some backup plans. 'If the weather means there is less food or more water, it's just more work on our end to make sure that the ko'ko's needs are met. If we have birds out in the wild and there's a drought, we'll put in water sources. If insect populations don't do so well, we can throw out food.'

In the worst-case scenario, a single typhoon could wipe out the rails from the wild. For that reason, there is an insurance population of about fifty Guam rails housed in zoos in the USA. Laura is confident that conservation on Guam can handle the changing climate: 'I feel it may be easier because we're not starting from a place where everything's good – we're already changing things. It's just a little step extra to think about the future.'

Thinking about the future is something she increasingly finds herself doing. She's aware that conservation on Guam is changing – it has moved from a focus on saving species to thinking about ecosystems. There have been many success stories, and Guam's wildlife is recovering. But the island's conservationists are yet to discuss their ultimate goals: 'Years down the road, could my project mess up someone else's? When we release birds such as the ko'ko', who predate other animals, could that be harming these species?'

Once again, discussions are needed about what we are ultimately trying to achieve with conservation.

Some conservation goals can be met even if a species becomes extinct in the wild. One common justification for conservation is the preservation of medicinal plants – every extinction, in theory, could deprive us of a new medicine. Likewise, wild plants could prove invaluable for our food production, providing genes which make our crops more resilient. With one in five plant species estimated to be threatened with extinction, we need to act fast if we are going to conserve the many species that could contribute to better crops or new medicines. We need to target many more species: scientists estimate that our crops have between 50,000 and 60,000 wild relatives. There is huge potential for these to help us create crops which are more resilient, nutritious and sustainable, but we can't possibly give each of them the same care that the Guam rail received. We can, however, save many of them from extinction by creating a backup collection of their seeds.

The first person to develop the idea of a seedbank was Russian scientist Dr Nikolai Vavilov, born the son of a wealthy merchant in 1887. Influenced by the work of plant geneticist Gregor Mendel, he became interested in plant breeding, and studied disease resistance in oats, wheats and barley. He is credited with identifying the areas where domesticated plants originated, and travelled in search of their relatives. His explorations led him throughout the world and, together with his staff he collected more than 250,000 seed samples for conservation in the Leningrad seedbank. He imagined Leningrad's future as a global food-bank, with seeds from the countries where agriculture originated helping to end hunger.

In many ways, Vavilov's work came at the right time as famine hit Russia between 1921 and 1922. Drought struck the country, and the wheat harvest was just half what it had been before the Civil War. Lenin recognised the need to improve Russian agriculture and prevent another hunger crisis, and Vavilov was elected by the new Soviet Union to a mission to collect seeds of wild crop relatives. He planned to use these to create frost-, drought- and disease-resistant crops, but there was no way to produce the immediate results the Soviet government wanted. When Stalin's move to collectivisation of private farms reduced crop yields and led to widespread famine, Vavilov was an obvious scapegoat. Stalin turned on him, and deemed his seed-gathering efforts to be bourgeois science. As a result, Vavilov came under surveillance by the Soviet secret service, and was considered suspect because of his family and ties to Western science.

He continued his work, but was arrested in 1940 while on a scientific expedition to the Ukraine. Although the exact charges are unclear, he was accused of being a traitor and a spy and was sentenced to death. This sentence was later changed to twenty years in a work camp, but that turned out to be the same thing. He died of starvation in January 1943 and was buried in a common prison grave. Vavilov's colleagues managed the seedbank once he was gone, but they were no more immune to Russia's food shortages than he was. They had to guard the collection from hungry rats and the famished population of Leningrad, and nine of them starved to death rather than eat the seeds.

Thankfully, the fate of the Leningrad Seed Bank has been transformed and it is now the N.I. Vavilov Institute of Plant Genetic Resources. Around the world new technologies are enhancing what can be stored in seedbanks, and collecting expeditions mean that we are safeguarding more and more of the world's plants. International travel may be easier than it was in Vavilov's day, but the remote locations and the skills needed to identify species mean that collecting the seeds of endangered plants is no mean feat. One high-profile example was a recent six-year mission by the Crop Wild Relatives Project to collect seeds from the plant species which had been identified as most likely to contribute to food security. Scientists from twenty-five countries travelled by foot, horse, jeep, boat and elephant to collect over 4,600 seed samples. Some of these were from very common species, enhancing the genetic diversity we preserve, whereas others were species at risk of extinction. Some of them were so rare that the trip may have come too late – the team had almost given up hope of finding a particular type of pea with edible tubers when an Italian researcher spotted a patch of unmistakable red flowers out of a train window.

Once the seeds have been collected, they must be stored somewhere both secure and accessible, and the seedbanks need meticulous records, as well as complex systems of backup power. No refrigeration system is infallible, though, and seedbanks are also vulnerable to more dramatic events such as natural disasters and war. The ultimate backup, therefore, is the Global Seed Vault, installed 1,300km north of the Arctic Circle, deep below ground in the permafrost on the island of Svalbard, midway between mainland Norway and the North Pole. Here, seeds are stored 100 metres inside the mountain, under layers of rock up to 60 metres thick. A cooling system keeps the temperature at a constant −18 °C, but the below-zero temperature of the permafrost means that seeds would be protected even if this failed.

The Global Seed Vault has already proved its worth, when the Syrian civil war forced the International Center for Agricultural Research in the Dry Areas (ICARDA) to relocate from Aleppo to Lebanon. The seedbank withdrew backups stored in Svalbard to grow replacements for seeds lost in the emergency move. ICARDA performs a vital role in protecting crop varieties and this work would othereise have been compromised.

For some plants, saving the seeds is not enough – there's a risk that treating seeds to remove microbes which could cause them to rot can also destroy those essential for the plant's growth. Symbiotic microbes are therefore conserved alongside seeds for species such as orchids, which can only grow with the help of particular fungi.

Elsewhere, protecting microbes isn't just a necessary add-on for seedbanks, it's a conservation project in its own right. Advancing technologies have allowed us to create the microbial equivalent of a seedbank. This effectively already exists in nature, with microorganisms preserved for millennia in ice and permafrost potentially being released in a warming climate. For humans the concept is much newer, and so far this has focused on microbes associated with the human body. The pioneering Global Microbiome Conservancy (GMbC), for example, stores stool samples from around the world. Their particular focus is on Indigenous and rural populations, whose gut microbes have been much less affected by antibiotics and modern diets. In this way, they can conserve microbial diversity that is being lost due to our industrial lifestyles.

This exciting project is at the forefront of ethical discussions about our microbes. It builds on the debate about ownership of human cells in the case of Henrietta Lacks. A cell line created from Lacks' tumour cells was used around the world for scientific research, without seeking her permission. A huge debate followed, and now ethical approval is sought from patients before use of their samples in research. At the GMbC, anyone who contributes to the project retains ownership of their samples. There remain difficult issues though, as people could one day be identified from the genetics of their gut microbes. What rights do you have to information about the bacterial cells within your gut? With scientists taking samples from rural volunteers overseas, this question is important and fascinating. It is intimately connected to ideas about what it means to be human. Where do you stop and other organisms start?

Questions about ownership have become even more complex with the arrival of new genetic technologies. As well as indefinitely storing faeces samples and isolated bacteria at −80 °C, the GMbC sequenced bacterial genomes so they can store genetic information electronically. This is far cheaper than storing a physical sample, and maybe it is just as good. Synthetic biology is rapidly advancing, and we can imagine a future where it is easy to reconstruct living organisms from their genetic sequence.

Is this what preventing extinction could look like − storing the digital instructions to recreate the organism? It has great potential, but brings us back to questions of why we want to prevent extinctions. A digitised code can't play the invaluable roles of microbes in the wild. Likewise, seedbanks around the world have huge potential to save both species and genes from extinction, protecting the diversity arising from evolution and from artificial selection by humans. However, if a plant species only exists as seeds in a bank, have we really prevented an extinction? If the seedbank is the end point, then few conservation goals would have been met. No seeds can fulfil the roles of the plants they can become − they can't store carbon, protect the soil or provide food for humans or animals. The seeds provide hope that these things will happen in the future, but their presence in a seedbank is just a step in the journey.

For wild crop relatives to contribute to a sustainable food supply, the seedbank is certainly a positive step. The ultimate objective is genetic information being incorporated into a crop, whether this is through traditional methods or the latest genome-editing technology. For this to happen, seeds must be available to scientists, and seedbanks help achieve this. For many species, the seedbank is only an unfortunate but necessary step to ensure they might one day thrive again in the wild. It's a sobering thought – what are the chances of this happening? These species may be on the ark, but will they ever get off it? There's a risk that we use seedbanks simply to delay the moment when we admit that something has gone forever. We can be hopeful, though, and while seeds remain in storage we can tackle the sustainability challenges that led to the species becoming extinct in the wild. It gives us future options for creating the world we want to see.

The constant genetic changes we see in nature lead us into a murky area related to extinction: what if a species takes on a new form rather than dying out? It's unclear when one species becomes another, and we risk getting tied up in debates about definitions (single-celled organisms don't obey neat rules of categorisation). As a result, we can lose sight of the real question: what are the impacts of those changes? And are genetic changes inherently wrong? These issues can come with both practical and cultural considerations, as the story of the American chestnut reveals. The extinction of this iconic tree may only be prevented by intentionally changing its DNA. Is this the right thing for us to do?

American chestnuts once grew across the Appalachian Mountains, from northern Mississippi to coastal Maine. Trees as tall as ten-storey buildings supported a major timber industry and were the main source of tannins for leather. At the dawn of the twentieth century, American newspapers showed train cars overflowing with chestnuts. City residents bought huge quantities, unaware that this delicacy would suddenly vanish. American chestnuts had dominated forests for 40 million years, but then disappeared within 40.

Their downfall began in 1904, when chestnut blight arrived on ornamental Japanese chestnut trees imported to furnish New York's expanding Bronx zoo. This fungus is first visible as small orange lumps on the bark, sometimes growing so they merge into one large patch. A sunken bulge of dead plant tissue, called a canker, then spreads to surround the trunk, and a toxin released by the fungus kills everything above. Chestnut blight spread rapidly, and by the 1950s large American chestnuts had all but disappeared and were declared functionally extinct. In total the fungus infected up to 4 billion chestnuts – a thousand for every person in America.

People were helpless to stop this loss, so attention turned to creating a resistant tree. Breeding programmes actually began in the 1920s, when American chestnuts were crossed with Asian species tolerant of blight. These attempts continued for decades, but all these early programmes failed. Efforts were revived in 1983, when the American chestnut was hybridised with the Chinese chestnut, its shrubbier, resistant cousin. So far, 70,000 hybrid trees have been grown and the aim is to fell all but the 600 most resistant. Early results suggest that, while the trees can't completely withstand the fungus, they may have enough resistance for reintroductions to work.

However, new technologies have allowed breeders to set their sights higher. By adding a wheat gene for an enzyme which breaks down the blight's toxin, they have created an American chestnut that is fully resistant to the blight. The genome of the genetically modified tree is over 99 per cent identical to the American chestnut, whereas the trees created through traditional breeding are about 94 per cent American and 6 per cent Chinese chestnut. The genetically modified trees do, however, come with regulatory hurdles that will delay their release into the wild, if indeed they ever make it, and at present are confined to carefully controlled plots. Each flower is contained within a mesh bag to ensure that no pollen escapes, and each developing nut is protected from squirrels by aluminium mesh. However, there are promising signs that a wider release could be appropriate. Tests have shown that the trees have no detectable effect on organisms such as bees and beneficial soil fungi, and that they produce nuts which are indistinguishable from those of native trees.

Does it matter that the new chestnuts have genes from other species? Such mixes occur regularly in the wild; the process of incorporating genes into a different species takes place 'naturally', without the help of humans. An example comes from another American tree – the south-western white pine. This is spreading northwards in response to climate change, a response that for some species will be the difference between survival and extinction. Its spread may be facilitated by locally adaptive genes which it gains through inter-breeding with the limber pine.

Cultural issues are also complex. Blight-resistant chestnut trees are currently being grown on field sites in the territory of the Six Nations of the Haudenosaunee Confederacy – one of the world's oldest participatory democracies. The 'people of the longhouse' see the law, society and nature as equal partners, yet have lost the rights to lands which they view as their mother. The Haudenosaunee don't all share the vision of the chestnut project, and many see it as a means of controlling nature. To some, all attempts to restore the forest go against the culture of not intervening – it's not the genetics that are the problem, it is the whole outlook. So, one part of deciding whether to restore this lost species to its former range is wide consultation, including with Indigenous communities whose voices have in the past been silenced.

Conservation also needs to consider the ecological impact of reintroducing the American chestnut, which goes well beyond whether or not the new trees are equivalent to their predecessors. The chestnut's loss had huge ecological impacts, but what will its restoration do? America's forests have changed and adapted, so planting the chestnut won't simply return things to the way they were before. It won't bring back the species that were lost alongside it. The blight also spelled the end for the phleophagan chestnut moth, for example, which unwittingly brought about its own demise by spreading the fungus. Much of the wildlife which the chestnut supported still survives, though, so its return could benefit species such as the wild turkey, blue jay and American red squirrel. The chestnut's fast growth and tolerance of poor soils also make it perfect for reforesting degraded land, such as the barrens left by surface mining. This would bring huge benefits, from carbon sequestration to recreation.

The story of the American chestnut highlights the message that no species are immune from rapid extinction, however common they are. It also illustrates what we can lose alongside a species, from livelihoods in the nut-crop industry to food and shelter for other wildlife. However, there's also a message of hope from the promise that we may again see American chestnuts thriving in the wild. With the emergence of new genetic approaches, such as CRISPR, other species may be able to make a similar comeback.

The scarlet honeycreeper in Hawaii is one species whose fortunes could be reversed with genome editing. This stunning bird is as bright as its name suggests, with a pink down-curved beak and scarlet body contrasting with jet-black wings and tail. Known locally as the i'iwi (*ee-EE-wee*), its populations are declining due to avian malaria and it has become largely restricted to higher elevations which are too cool for mosquitoes. With the climate changing, such areas will no doubt shrink, meaning the species is threatened with extinction. Could we use genome editing to create malaria-resistant honeycreepers and release them in Hawaiian rainforests? Scientists are only at the stage of assessing the potential of this 'facilitated adaptation' for scarlet honeycreepers, which comes with technical challenges. However, if it works it could be a long-term solution.

Some scientists are going one step further, with attempts to recreate species that have already gone extinct, such as the passenger pigeon. In 1901, the last wild passenger pigeons were shot – just decades after their population had numbered in the billions. Some skins were preserved, though, which has allowed their genome to be sequenced. This makes it possible to edit the genome of the living band-tailed pigeon to create offspring close to their extinct cousin. Could modern genetic technologies mean we see passenger pigeons once again in the wild? Again, there are technical hurdles to overcome, but such possibilities mean it is time to seriously examine the social and ecological effects of recreating extinct species, and using genetic technologies to prevent and reverse extinctions.

As our knowledge expands, we can make better decisions about how to prevent extinctions and about which species we should prioritise. We can't save everything – we definitely have choices to make – but we can reverse our current trajectory to ensure that more species thrive, ourselves included. There will be costs and trade-offs, but the potential gains are huge. If our metaphorical ark sinks, we are going down with the ship.

CHAPTER 5

ANIMAL WELFARE

Floreana mockingbird vs rats (and a debate on culling, rewilding and kangaroos)

Floreana's elegant mockingbird is the size of a slender blackbird, with a narrow down-curved beak perfect for feeding on prickly pear cactuses. It was the smallest of the four endemic mockingbird species Darwin found in Galápagos in 1835, and has long been at the greatest risk of extinction. Darwin recorded mockingbirds as being common on the island of Floreana, but that changed in just a few decades, and the last reported sighting on the main island was in 1868. The bird is now confined to two islets: Champion and Gardner-by-Floreana. These tiny landmasses are vital refuges for the Floreana mockingbird, but they are so small that together they house fewer than 500 individuals. The population fluctuates dramatically due to extreme weather events, bringing them perilously close to extinction.

Floreana is the smallest of the four inhabited Galápagos Islands, about 173 square kilometres in all. It has a curious history of human settlement, as home to the Galápagos's first resident, an Irishman named Patrick Watkins who was marooned there in 1807. He survived by growing vegetables, which he traded with passing ships for rum. He is said to have enticed four other sailors to remain by getting them drunk and eventually the five managed to steal a boat from a visiting ship in 1809 and rowed off to mainland Ecuador.

The next settlement on Floreana was intentional: a colony set up in 1832 to house soldiers who had taken part in a failed coup in Ecuador. The lack of fresh water meant the settlement failed, but in the twenty years it existed the island ecology was severely impacted. This included a rapid decline, and then extinction, of giant tortoises, which the settlers sold live and as oil. They also felled forests to make way for farmland and introduced animals such as donkeys, goats, pigs and cattle, all of which damaged the fragile habitat. But the biggest problem was the stowaways. Rats and mice arrived when ships docked on the islands and spread across Floreana. Birds' eggs and nestlings make a nutritious meal for the introduced rodents and, two centuries on, the results have been disastrous.

Originating in Asia, black rats have literally colonised the world, first alongside the Romans and later with European colonists. Like humans, rats are omnivores, and their diet of seeds, fruits, leaves, insects, eggs and other animals is one reason they have proved so adaptable. It is also why they have decimated populations of native species.

Sometimes people try to keep rodent numbers down with pet cats, but this simply adds to the problem for wildlife – the domestic cat hasn't lost its predatory instinct. The problem of introduced cats and rodents has been particularly stark on islands, where species often evolved in the absence of native predators. The Floreana mockingbird, for example, adapted to spend a lot of time on the ground. It is therefore easy prey, so can't survive anywhere that predators roam. Another casualty is the critically endangered Galápagos petrel, half of whose global population breed on Floreana. Like many seabirds, the petrels build their nests on the ground, so their fluffy, flightless chicks are at great risk. Park wardens now regularly control cats and rats around the nests, so the petrel population has increased, but it's an ongoing struggle.

And it isn't just the birds that are in trouble. Galápagos racers, which became a viral trend after the BBC TV series *Planet Earth II* showed footage of a group of the snakes hunting in a pack, are in trouble from predators. Research following their TV appearance discovered that there are nine different species of racer in the Galápagos, each with a very small and highly vulnerable distribution. Smaller species are threatened too. Rats and mice prey on native snails, and at least thirteen of the twenty land snail species unique to Floreana are threatened with extinction. Plants can likewise be affected – prickly pear cactuses are prone to toppling when rats eat the stems.

It's clear that, for as long as cats, mice and rats thrive on Floreana, other wildlife will suffer. Plants, animals and eggs will be eaten, and the mockingbird will remain confined to its two tiny islets. This doesn't have to be so on Galápagos. Pinzón Island has been rat-free since 2012, an achievement which that ended almost 50 years of rearing giant tortoise hatchlings by hand and releasing them only when they were old enough not to be predated by rats. In 2014, hatchlings survived in the wild on Pinzón for the first time since the 1800s. This was a triumph for the partners involved in the eradication, which include the Galápagos National Park Directorate and the charity Island Conservation.

Gregg Howald was an ecotoxicologist on this project, and spent two decades working with Island Conservation to restore island ecosystems. He has seen success stories from around the world, and witnessed the power of eradications. 'The evidence is crystal clear,' he says. 'We know that when invasive mammals are removed from islands the ecosystem will respond. Everything from plants to insects and birds can recover.' Island Conservation and their partners have therefore set themselves the challenge of eradicating introduced predators from Floreana. On an island ten times the size of Pinzón, this is a bold project. Every last rat, mouse and feral cat will need to be removed and then additional work carried out to restore the ecosystem.

Floreana's native vegetation has already benefited from eradication of introduced species, after 1,561 goats and 380 donkeys were removed in 2006–09. The practicalities of eradicating rats and mice are very different, though. Whereas larger animals can be hunted from helicopters or with dogs on the ground, targeting individual rats is impossible. As Gregg explains: 'The only way we know to eradicate rats and mice from an island larger than the size of a football field is to use rodenticides. You don't know where the animals are, so you have to put bait into every potential territory.'

Given Floreana's size and rugged terrain, some areas can only be targeted by dropping the poisoned bait from a spreader bucket suspended below a helicopter. Initially this will target rodents, then in a second stage planes will drop sausages containing poisons specifically directed at cats. The final surviving feral cats will be trapped or hunted with dogs and shot, and all pets will be neutered.

Spreading poison over such large areas inevitably affects non-target species. Gulls and birds of prey are poisoned when they eat dead and dying rodents, and other native species are at risk from eating the bait directly. Much of Gregg's time is therefore spent tackling the question of how to deliver poison to the target species without causing problems for other animals. He explains the solutions: 'For some species all we can do is take a small number into captivity then release them when there is no longer a risk. Other species are more abundant, so we accept the loss of a small number because the population will bounce back.'

At a population level, this solves the problem – animals lost due to poisoning will be replaced. But should this satisfy us? Individual animals will suffer pain and death from poison, and those taken into captivity will be stressed. Killing animals for conservation can understandably trigger heated arguments. 'People are often very worried about welfare impacts on native species, and are sensitive to the issue of feral cats,' Gregg says. 'There's a high reluctance to manage cats and it gets political very quickly. Rat welfare, on the other hand, is rarely considered.'

An animal's capacity to suffer isn't related to how rare it is, how appealing we find it or whether we view it as native. Branding certain species as pests doesn't mean we can wash our hands of the moral implications of killing them. It's clear that the rats and mice on Floreana will suffer, sometimes taking days to die. The poison to be used is brodifacoum, an anticoagulant related to the blood-thinning medication warfarin that inhibits production of vitamin K. Any animal that ingests even a small dose will slowly bleed out internally.

The eradication therefore has grim consequences, for rodents, cats and non-target species. At the same time, eradications can also bring to an end long-term poisoning programmes. And ridding Floreana of its introduced predators will prevent the ongoing harms they cause to other animals. Gregg is acutely aware of the suffering caused by introduced animals: 'Predation by a cat, for example, incurs the same horror as someone hunting an animal, often worse. And they are only there because of us.'

Rodents can cause painful deaths for the birds, which often have minimal defence mechanisms. BirdLife International host a YouTube video of a fluffy albatross chick being eaten alive by mice on Gough Island, in the Atlantic Ocean. As Gregg explains, there's good reason why the video starts with a graphic content warning: 'You can see the mice jumping at the chick's eyes or on their back, and systematically biting and biting until the birds eventually die. It's horrific.' In this way, mice can cause the death of chicks that are fifty times their size. An estimated 2 million fewer seabird chicks fledge each year as a result.

Eradication campaigns may attract more attention than long-term planning but they don't necessarily cause more deaths, and whether or not we eradicate the predators animals will suffer. We shouldn't be afraid to acknowledge this. Gregg is a strong believer that we need to be open and honest about the trade-off. 'Nobody wants to see pain and suffering in a human or an animal,' he says. 'Personally, I feel bad for the animals that are lethally removed from an island – I know what a poison can do to an individual animal. But I have to frame it in a cost–benefit analysis. A lot of the projects I work on focus on endangered birds or reptiles. If we don't do the eradication, we are going to lose these species.'

As we saw earlier, there is no intrinsic reason to prevent extinction, no trump card which makes it a forgone conclusion that it is worth killing animals to keep a species around for a little longer. There is also no correct historical baseline for how common a species should be, no point at which it is automatically right to start killing animals of one species to increase the numbers of another. However, there can be many benefits to protecting species and populations, and there is plenty of evidence that eradications of introduced species on islands are an effective tool for species conservation. An eradication programme on the Canadian island of Langara, for example, was followed by an increase in ancient murrelets, a relative of the puffin. On Anacapa, one of California's Channel Islands, Scripps's murrelets have rebounded, and ashy storm petrels have returned.

Island eradications can also bring great benefits in the surrounding ocean. Seabirds catching fish bring nutrients back to land, and these enhance the productivity of island wildlife. Some of these nutrients flow back into the sea, which has a positive effect on the ecosystems near the shore. Coral reefs in particular benefit, and scientists have found that the coral reefs next to rat-free islands thrive as a result of the nitrogen reaching them via seabird guano. The benefits spread widely – herbivorous fish, for example, are more common and grow faster on reefs next to islands with no rats.

The benefits of rat and cat eradication on Floreana are expected to be wide-ranging. Birds such as the medium tree finch and the Galápagos martin should recover of their own accord and with some help other species should flourish too. Floreana mockingbirds will be reintroduced, as will the Floreana giant tortoise. This island endemic was thought to have gone extinct by 1850, partly due to rats eating tortoise eggs and hatchlings. However, the species was recently rediscovered by comparing the genetics of museum specimens with wild tortoises on neighbouring Isabela Island. It seems that Floreana tortoises were translocated to Isabela by whalers, and their hybrid descendants live on. A breeding programme of twenty tortoises began in 2017, so Floreana can be repopulated with tortoises that are genetically similar to their ancestors. Weighing up to 320 kilograms and more than 1.2 metres long, the Floreana giant tortoise performs important roles in an island ecosystem, such as seed dispersal, soil disturbance and grazing. Its return should have a positive effect on the habitat.

The eradications will also benefit Floreana's people, and this is the first time in Galápagos that a community will be directly involved throughout the conservation programme, from planning to evaluation. Most of the population are subsistence farmers and fishers, and the conservation goals can bring wider benefits for them. Agricultural yields are likely to increase once rats and mice are eliminated and the strategies used to protect domestic animals from poison will also have long-term benefits. In particular, facilities on farms are being improved so that animals can be kept safe when bait is dropped. 'This is actually increasing productivity,' Gregg says. 'For example, better chicken coops mean people are collecting more eggs – they haven't been laid behind a tree or shrub where nobody checks.' The new enclosures for farm animals in turn benefit wildlife. Native birds are susceptible to diseases carried by chickens, and improved animal husbandry reduces this threat.

For the majority of conservationists, all these benefits justify the killing of rats, cats and other species that have wreaked havoc on islands. In other situations we kill animals for far lesser gains and, as Gregg points out, 'The number of animals killed for conservation palls in relation to what's going on globally. The volume of rodenticide that conservationists use is a pittance compared to what's going on in cities, on farmland and in the homes of regular citizens.' Not everyone agrees, but it reminds anyone who is opposed to killing for conservation to take a look at the impact of their own lives first.

It is a sobering thought for each of us to consider how many animals have died in our names. Even though I eat very little meat, my list is alarmingly long. My food is a major reason for this, and not just the meat. Milk production, for example, has no need for the male calves, so they are not reared to adulthood. The same is true of egg-laying chickens. Rats and mice have been trapped or poisoned in my house and garden, on farms that grow my food and in parks and gardens that I visit. I benefit from medications, including vaccinations, that were tested on animals, and own leather shoes. I've been in a vehicle which killed an animal, and I have taken ill pets to the vets to be euthanased. And that's not to mention more indirect deaths – the mining of precious metals to make my many electrical goods will have displaced animals through habitat destruction. This acknowledgement provides essential context to discussions about whether it is acceptable to kill animals for conservation. It's also a reminder that we should not simply accept the deaths that come with our lifestyles.

Killing animals is a key tool in conservation, and is essential to protecting species such as Floreana mockingbirds from extinction. However, the increasing concern for animal lives and welfare in wider society means we can't ignore the ethical issues. There are two general approaches to the debate and it helps to understand the difference between animal rights and animal welfare. To an animal rights supporter, it is always wrong to take an animal's life, regardless of any benefits that would bring. That's certainly how we see people. The Universal Declaration of Human Rights includes the right to life, liberty and security of person. With animals, we tend to see things differently. It wouldn't be justifiable to do medical experiments on people without consent, whatever the benefits, yet most of us accept this for animals. Animal rights supporters, on the other hand, believe we should extend some basic rights to animals. For example, they may believe that it is wrong to kill an animal other than in self-defence (as with humans).

We can, however, view animal ethics through a different lens. Rather than focusing on animal rights, we can instead concentrate on welfare, and think about how to reduce suffering. The different points of view are well illustrated by the vegan fox thought experiment:

> *Imagine that, just as you are settling down for a countryside picnic, you see a fox about to catch and eat a rabbit. As foxes are omnivores, your vegan meal would be perfectly good for the fox. There are two choices here: (a) eat a vegan meal and let the fox catch the rabbit, (b) eat the rabbit and give the fox a vegan meal.*

To an animal rights advocate, the second answer is out because you have killed an animal. However, from the rabbit's perspective it could be vastly preferable to (a) because you're likely to give it a less painful end. Therefore, if you are concerned for animal welfare, killing the rabbit is arguably a better choice. Concern for rights or welfare can therefore cause people to reach completely different conclusions about killing for conservation.

The answer isn't clear-cut even for someone making a judgement based on welfare. By choosing (b) over (a), we may reduce the rabbit's suffering, but we deprive the fox of natural behaviours. This is a huge challenge for animal welfarists – how can we know what an animal wants? How can we compare physical pain to an animal's freedom? Such trade-offs are hard enough to balance for ourselves, let alone for a wild animal.

Some people would rule out giving a vegan meal to the fox because they believe that humans shouldn't interfere with nature. The fox should be free to kill the rabbit. And choosing not to prevent suffering is arguably different to inflicting it. Of course, the reality is that we already have hugely altered nature, which is why cats are catching birds in Galápagos. Given the negative impact we have had on nature, it would seem to be ironic if we decided positive interventions would be wrong. However, some people would maintain that the appropriate relationship between humans and individual animals is for us not to interfere.

The hypothetical situation of the fox and the rabbit isn't designed to argue for a correct response, but to reveal that there are very different ways of approaching our responsibilities towards animals, each of them legitimate. Interesting discussions also arise if we substituted the rabbit and the fox for different species. We might see things very differently if it was a Venus flytrap about to catch a fly, for example. If the Venus flytrap can't feel the thrill of the chase, and the fly feels no fear, the decision has different consequences.

This leads us into the murky controversy about which creatures have any capacity to feel pleasure or pain. Scientists and philosophers debating this issue usually use the word 'sentient' to describe an individual with some kind of capacity to feel. However, there is no one single definition of what sentience means. It may be consciousness, self-awareness or the capacity to feel physical pain. We could ask: 'does this animal experience anything?' If so, then it is sentient.

Firm answers to such questions of sentience are a long way off. But current evidence suggests that all vertebrates are sentient, while invertebrates are not, with the exception of squids and octopuses. However, some scientists argue that we haven't yet confirmed that fish can feel pain. And an increasing number of scientists believe that some invertebrates such as snails, insects and spiders can feel pain. We are constantly making discoveries such as new types of pain receptors, so we may simply lack the tools to demonstrate widespread sentience. It seems likely, however, that even if an insect has some form of sentience it can't experience the same level of suffering that a mammal can. This is one reason that the thorny issue of wild animal welfare focuses almost exclusively on the vertebrates. Maybe one day it will be possible to make ethical decisions based on knowledge of what it is like to be a bee, but for now it's challenging enough to make decisions for the large animals we are familiar with.

In Charles Causley's poem 'One Day at a Perranporth Pet Shop', a lady buys a budgerigar and promises it a life of luxury. However, the bird is so horrified by the prospect of cream cakes and satin sheets that it flees to its ancestral home in Australia. At the end of the poem Causley invites us to agree with the conclusion that it is better to 'eat frugal and free in a far-distant tree, than down all the wrong diet in jail'.

This light-hearted poem about a budgerigar held captive in Cornwall raises a pertinent question about life's priorities. It's a complex one, though – how frugal, how free? What freedoms would you be willing to sacrifice in return for the budgie's whiskies and sodas? It can be hard enough to predict what answers another human will give to this kind of question, let alone a member of a different species.

Science continues to tackle questions about what is best for animal welfare. We can monitor an animal's cortisol levels to see how stressed they are, we can observe their behaviour and measure their pain. But it is very hard to maximise their welfare without ever feeling what it is like to be an animal. For this reason, some people advocate leaving animals to pursue their own lives, prioritising freedom over comfort. However, we shouldn't automatically assume that freedom is best for welfare – many domestic animals benefit from avoiding some of the suffering experienced in the wild, such as hunger and thirst. Just as I benefit from certain ways my freedom is restricted by law, animals may benefit from ways we restrict their freedom. My life is certainly free from some of the suffering I would encounter if I led the lives of my hunter-gatherer ancestors, and the same is true of captive animals receiving food, shelter and veterinary care.

These different views about welfare are at the heart of ongoing clashes between protesters and land managers at Oostvaardersplassen, just north of Amsterdam. This nature reserve was created in 1968 when an inland sea, Lake IJssel, was drained to make way for development. However, the development never happened and the site was abandoned, allowing a wetland to form spontaneously. Migratory geese colonised the area, generating a new habitat rich in birdlife. This sparked an idea for the charismatic and controversial conservationist Frans Vera, leading him to believe that the standard theory of what Europe used to look like was wrong. Instead of forest, he concluded that it was an ever-changing mixture of forest and pasture. Before humans modified the landscape, large herbivores maintained areas of open grassland, just like the geese do. Grazing prevented trees from dominating, and herbivore numbers were kept in check by predators such as wolves and lynx.

This insight prompted Vera to begin an ambitious restoration experiment in the early 1980s. His vision was to return lost ecological processes to Oostvaardersplassen. He wasn't interested in the intensive management of modern conservation, but in the self-managed ecosystems that humans encountered when they first arrived in Europe. Rather than counting how many species of birds made their home in Oostvaardersplassen, he was concerned with the levels of predation, grazing and decomposition. The question was how to achieve this, given that past ecosystems were shaped by herbivores and predators that are now extinct. He therefore sourced thirty-two Heck cattle (which we met in the first chapter) and eighteen Konik ponies, which were 'back bred' in the same way to resemble their wild ancestors. His hope was that the cattle and ponies would return to their wild natures and evolve to lose their domestic characteristics.

The cattle and ponies were joined by a release of forty red deer in 1992 and, as Vera had hoped, they thrived. By 2016 there was a combined population of around 5,300 animals. The growth in numbers, however, was by no means steady, as harsh winters reduced the number of animals the reserve could sustain. From Vera's perspective, this was integral to creating the landscape he envisaged. Each winter a proportion of the population dies off, so come the summer there aren't enough animals to graze the whole reserve equally. This allows areas of denser vegetation to grow, creating a patchwork of habitats where different species can thrive. However, the winter die-off can be thousands of animals, which from an animal welfare perspective can be a tragedy.

The reserve is next to a railway line, and the sight of emaciated animals and decaying carcasses caused shock and outrage when they appeared on television and online. An 'Anti Oostvaardersplassen' Facebook page shows a video, taken during the harsh 2009–10 winter, of a deer sprawled on its side in some flood water, still twitching as it makes feeble attempts to lift its head for air. People have responded in different ways. Some threatened ecologists and rangers on social media, others threw bales of hay over the reserve fence to feed starving animals, and a Dutch animal welfare organisation launched a court case to demand an 'end to experimentation'.

The court case turned into a lengthy legal dispute, with the managers eventually winning the right to continue without intervention. However, the whole episode was understandably a public relations disaster, and a compromise was needed. Following the recommendations of an expert panel, it was agreed that new forest and marsh areas would be created for increased shelter, and the animal population would be controlled. Ponies were captured and transported to new homes, and rangers patrolled the reserve and shot animals they believed wouldn't survive the winter.

The debate, however, was far from over, and led to repeated court cases, with some groups arguing in favour of culls to prevent what they see as overgrazing. For a time, those who advocated culling appeared to be winning, and during the winter of 2018–19 around 1,800 of the reserve's 2,300 red deer were shot. Likewise, cattle and ponies were shot or taken away for slaughter. However, many nature organisations lodged an appeal against this culling. Frans Vera was one of the opponents, and the legal challenges he supported prompted the court to pause the cull in 2019. As of 2021, ongoing court battles seem stuck in a stalemate, and only weak and older animals are shot in the winter.

The ideological debates around the culls centre on different perspectives about animal welfare. Some people object to the animals' deaths, and others object to the animals' lives. The people concerned with animal welfare would argue that the animals' lives should be more carefully controlled to ensure their comfort. Seeing cattle and horses in human protection, not left to fend for themselves, is certainly what we're used to. Dutch Olympic gold medal rider Anky van Grunsven, for example, has condemned Oostvaardersplassen as animal abuse, an outlook which is anathema to Frans Vera. To him, any idea of wild animal welfare based on the standards for livestock is misguided: 'With the disappearance of wild cattle and horses, an understanding of their welfare has disappeared, and, for these species, has been redefined according to experiences with domesticated animals on farms. The fact that wild-living cattle and horses in the Oostvaardersplassen had a completely free life with a natural social order, that the calves and foals stay with their mother, and have a natural social order like that of other large bovine ungulates and equids living in the wild, did not seem to matter. This aspect of their freedom is forgotten or ignored.'

Wildness comes at a cost. We've all watched as ducklings disappear one by one so only a few make it to adulthood, and seen penguin chicks freeze to death in wildlife documentaries. Most of the young birds that survive long enough to attempt an autumn migration will never return to breed the following spring, and predators cause fear, pain and death. Part of the conflict is that Vera and van Grunsven have different views about whether Oostvaardersplassen's animals are domestic or wild. It is a far from clear-cut issue. The cattle and ponies have domestic ancestry but have been specifically bred to reverse some of the effects. The animals are constrained in a 56 square-kilometre fenced reserve, which prevents them from migrating in search of food, yet gives them a freedom to roam that is unimaginable for livestock and pets. They have little contact with humans, yet humans are masterminding their fate. Their condition is now regularly assessed by a vet, an important part of domestic animal welfare that rarely features in a wild animal's life. Debates about domesticity risk missing the point, though – categorisation of wild and domestic is of no relevance to an animal.

Some people promote the introduction of predators to control populations, which arguably means humans are not responsible for each death. However, being killed by a predator is more painful and terrifying than being shot by a trained marksman, and it won't prevent animals from starving. In the Serengeti, for example, more animals die from malnutrition than predation. This raises the question of whether we should intervene in less managed habitats. We may not see ourselves as responsible for animals in the Serengeti as we do those at Oostvaardersplassen, but if we have the power to reduce their suffering perhaps we should consider it.

There's no easy definition of a wild animal, as we saw in chapter 2. From the birds that eat my garden peanuts to the hand-reared tortoise hatchlings released in Galápagos, there is ambiguity everywhere. This is another argument for moving beyond simply asking 'are we responsible for these animals' to questioning whether we can do anything to reduce their suffering. It's a big question.

Despite the controversy surrounding Oostvaardersplassen, Frans Vera's vision for a landscape shaped by free-ranging herbivores has captured imaginations. Other reserves have followed his lead, though generally without the concerns about animal suffering. One high profile example is Knepp Castle Estate, a 14-square-kilometre rewilding project in southern England. This former farm is now sustained by eco-tourism, with an important extra source of income coming from the estate's organic meat. High welfare standards are one reason behind the meat's premium price, revealing a stark contrast between Knepp and Oostvaardersplassen. Similar conservation strategies can, it seems, be accompanied by very different attitudes to managing animals.

Knepp is home to hundreds of grazing animals, including English longhorn cattle, Exmoor ponies, Tamworth pigs and red and fallow deer. The longhorn cattle are iconic, with their huge horns and mottled russet and white fur. These cows have been bred for docility, and their hardiness makes them ideal for this kind of unrestricted living – in bad weather, the cattle often choose to shelter in woods or hollows rather than the empty barns. They're free to roam the estate, forming small herds and passing knowledge from one generation to the next. They are closely monitored, though, and are rounded up for tuberculosis tests and for slaughter. Even this is done with care for their welfare. The estate's stockman watched online videos of an American 'cow whisperer' who teachers ranchers to round up cows on foot, and adapted his teachings to create a form of low-stress stockmanship that works in the complex landscape of the Knepp Estate.

Knepp's stockman monitors the animals' condition, as well as the stock density, which is kept low enough that there's no danger of animals starving over the winter. Where Vera was happy for the habitat to be completely shaped by the herbivores, Knepp had some preconception of what they wanted to achieve. They share Vera's willingness to embrace uncertainty, but only within defined limits.

Vera's desire for surprising and unprecedented events is in complete contrast to mainstream thinking in European conservation. He didn't start out with targets, action plans or a desired habitat, but with a trust in nature. After all, nature has never been in a static equilibrium, so why should we expect it to stay constant now? The results have been positive in many ways, with an enhancement of ecological services such as carbon storage and flood control. However, to many conservationists this isn't enough to herald Oostvaardersplassen as a success – critics argue that the climax vegetation was forest, or that by not having specific targets Oostvaardersplassen doesn't support the species we should be focusing on. Although Oostvaardersplassen has become home to rare invertebrates and birds such as bearded tits, bitterns and white-tailed eagles, this is a happy accident. To many people, these rare species should be the focus of conservation and so the land should be managed specifically for them. If we want to protect rare species, we can't rely on surprises.

In conservation terms, Knepp takes a middle ground between Oostvaardersplassen and traditional target-led conservation. They manage their herbivore populations to ensure there aren't so many animals that the estate becomes grassland nor so few that it becomes closed-canopy forest. Exactly what the stocking density should be within this range is open for debate, and Knepp is aware that there is no way of defining an 'optimum' level of grazing. Essex skipper butterflies need long grass, for example, but Dor beetles thrive in tightly grazed 'lawns'. As the estate continues its journey from commercial farm to wild space, there will no doubt be ongoing discussions about what level of management should take place while still allowing nature to set its own course.

This attitude of non-interference is central to the current trend for rewilding, something we will explore in the following chapter. However, as we consider the tensions between rewilding and land management, we must remember that an animal's suffering is not affected by conservation goals or outlooks. Pain doesn't become any less because their habitat resembles a past state, or if the reserve is home to species that conservationists have deemed to be important. An ideology of leaving animals and habitats free from human intervention brings exciting possibilities, but Oostvaardersplassen has shown that this won't necessarily be best for animal welfare.

Knepp, on the other hand, reveals that it is possible to have the surprises which come from allowing herbivores to shape the environment without the mass starvation that comes from leaving populations unchecked. Perhaps this is a middle way, allowing animals to live lives of freedom while their deaths are more controlled. There are no simple answers, though, and both Knepp and Oostvaardersplassen are helpful reminders of how concern for animal welfare needs to be considered alongside conservation debates about what rewilding can achieve.

The one situation where it is indisputable that humans have taken responsibility for welfare is when animals are taken into captivity. This happens for all sorts of reasons, including for our entertainment, to care for injured animals and for conservation. Bringing animals into captivity has contributed to some conservation triumphs, as the story of the Guam rail showed us. However, it can come at a cost for the individual animals, causing stress and sometimes even death. Personally, I've only witnessed this on rare occasions, but when I have it has been dramatic. A highlight of doing fieldwork in Gibraltar was watching the spring bird migration, as black kites, honey buzzards and short-toed eagles streamed over from Morocco. Among the other birds in the mix were griffon vultures, which attract the attention of the yellow-legged gulls.

As carrion feeders, the vultures pose no threat to nests or chicks, but the gulls see these huge birds as dangerous. Flocks of angry gulls therefore mob the vultures, sometimes hitting their backs, and in extreme cases drive them down into the sea. My friends at the tiny raptor rescue centre would be watching, and head out in a speedboat to save the vulture. If they made it in time, they would bring the vulture back wrapped in a blanket, and keep it at the rescue centre while it regained its strength. Soon it would be flying at the wall of its cage, agitating to be let free. And it would be. In this instance, the birds' lives were saved by a few hours in captivity, but in other situations healthy animals may remain in zoos or aquariums for the rest of their lives.

The debate is particularly emotive when we deal with cetaceans (whales, dolphins and porpoises). These highly intelligent mammals often live in social groups and range over large areas of rivers and oceans, yet have been routinely kept alone in tiny pools. Awareness was raised by Keiko the killer whale, who played the lead role in the 1993 film *Free Willy*. In the fourteen years between Keiko's capture and the film's release, he had performed in amusement parks, been transported across continents, developed skin lesions due to poor health and been bullied by older whales. He was kept in a shallow pool in Mexico that had been designed for smaller dolphins, whereas killer whales in the wild regularly dive to depths of around 75 metres and travel for thousands of kilometres in large pods. Unlike Willy in the film, Keiko's story doesn't have a happy ending. He was indeed released, following a public reaction sparked by the film, but never integrated into a pod or adapted to life in the ocean. He died in 2003, apparently of pneumonia.

Since then, our understanding of how to care for whales and dolphins has improved, as have welfare standards. However, there are still serious concerns about keeping cetaceans in captivity, making it a cause of great controversy. It's therefore no surprise that the 2018 announcement of an international workshop on *ex situ* options for cetacean conservation was met with protests from anti-captivity campaigners. The International Union for the Conservation of Nature (IUCN) wasn't deterred, however, and the workshop brought together biologists, conservationists, vets and representatives from zoos and aquariums.

Their focus was whether *ex situ* options could help the conservation of dolphin and porpoise species that are facing extinction. *Ex situ* was interpreted quite broadly, and included scenarios such as confining Yangtze finless porpoises to semi-natural oxbow lakes along the Yangtze River. Like so many cetaceans, the finless porpoise faces the major threat of becoming entangled in gillnets – curtains of netting that hang in the water. Gillnets provide food and livelihoods for communities but have been devastating for species that are caught alongside the fish, such as dolphins, sea turtles, seals, sea lions, seabirds and sharks. The problem can be reduced with solutions such as acoustic alarms to deter marine mammals, but these aren't effective for all species and are beyond the means of small-scale fishers.

Gillnets have played a big role in the decline of the critically endangered vaquita, which is found only in the upper Gulf of California in Mexico. These porpoises have distinctive facial markings, with darker grey around their eyes and lips, and grow to just 1.5 metres long. Attempts to halt their drastic decline have included removing lost gillnets and establishing a refuge area where gillnets are banned. Sadly these interventions have failed, partly because gillnet bans aren't consistently enforced. In 2017, with the estimated population dipping perilously low, it was decided to bring as many vaquitas into captivity as possible. A team of around ninety people from nine countries was assembled, but they quickly faced a setback when the first vaquita they caught became stressed. They released the young female later the same day. A second vaquita seemed to adjust well to a confined environment but died of capture myopathy, a condition associated with stress. The capture attempts were immediately halted, and the remaining vaquitas were left to take their chances in the wild. An estimated ten vaquitas remain off the coast of Mexico, so their extinction appears imminent.

The tragic story of the vaquita raises difficult questions about how, when and if to intervene. Deaths are to be expected in the early stages of handling a species which has never before been kept in captivity, so if *ex situ* programmes are going to work they need to begin when numbers are higher. This gives us time to learn how to care for the species and to ensure genetic diversity in the captive population.

A better strategy is, of course, to tackle issues such as gill-nets so that survival is possible in the wild. This has many benefits for both population numbers and welfare, and some people fear that conserving a population in captivity will detract from this main cause. However, the reality is that the issue of gillnets hasn't been solved anywhere in the world, and for species such as the vaquita captivity may be the best (and perhaps only) chance of avoiding extinction.

At the other end of the spectrum, animal populations that are very high or rising rapidly raise dilemmas related to welfare. The issues become particularly thorny when the animals concerned are killed, and disagreements can come from both the manner of killing and its motivation. The debates are made more complex by the fact that a single animal may be killed for multiple reasons – a sport hunter can kill a 'pest' animal, then sell the meat, for example. Each animal's death can therefore be embroiled in multiple debates. Is pleasure a legitimate excuse for killing? What counts as a pest? Does wild meat have a higher welfare standard than meat from livestock? Is it more sustainable?

All these issues can become entwined, and nowhere more so than in Australia's culling of kangaroos. The killing of Australia's national animal emblem divides the nation. Vocal opponents denounce the culls as an inhumane massacre, while supporters see them as a traditional aspect of rural life. After all, Aboriginal Australians have been hunting kangaroos for thousands of years, often making full use of the carcass: meat is eaten, teeth are used as needles and sinews from the tail are used as thread. Hunting, however, changed and intensified with the arrival of European settlers, who were quick to define kangaroos as pests that damaged gardens and competed with livestock for grazing. Between 1877 and 1907, around 8 million kangaroos and wallabies were presented to the Queensland Government alone for bounty money, pushing some species to the brink of extinction. Times have changed since then and kangaroos can now only be shot under license, ensuring that the harvest is sustainable.

Kangaroos are killed in different situations, ranging from farmers shooting 'pest' kangaroos on their land to professional marksmen taking part in commercial hunts for meat and leather. The line is blurred, though: some farmers allow licensed shooters on their land and view the commercial kangaroo industry as self-funding pest control. The definition of 'pest' is decidedly subjective in both farming and conservation, and opinion is divided on whether kangaroo culls benefit or harm the wider environment. One controversy is that farmers control dingoes to protect livestock, and this can allow kangaroo populations to grow. In high densities, kangaroos can reduce the diversity of plants, affecting habitat quality as well as the populations of other animals. They're also a problem for people, competing with livestock for food, damaging fences and causing car accidents.

However, at lower population densities, kangaroos play an important role as ecosystem engineers. Some conservationists argue that a commercial kangaroo industry could be the motivation we need to manage land to suit the kangaroo. Australia has been degraded by livestock ever since Europeans arrived, and replacing livestock with sustainable harvest of wild kangaroo meat could be a way to restore the rangelands. However, more evidence is needed to answer the question of whether this is a viable solution for habitat protection. In fact, we're short on data about the ecological impact of kangaroo culls in general, and that's before we've thrown welfare debates into the mix.

Just like in Oostvaardersplassen, there are animal welfare arguments both for and against culling. In good years, kangaroo populations boom, just as herbivore numbers do in Oostvaardersplassen. Drought years bring a bust, in which kangaroos starve and the land is degraded by overgrazing. Shooting kangaroos can therefore prevent deaths from starvation, although it doesn't mean the end is painless. Some animals are mis-shot, and the break-up of social groups has consequences for the animals left behind. Even those who are found may not come to an easy end: shooters are legally allowed to crush the heads of joeys with a steel water pipe.

The fate of joeys plays a prominent role in campaigns against supermarkets selling kangaroo meat and sports companies using kangaroo leather to make football boots. The campaigners can boast some successes. A 2016 import ban ended the thriving kangaroo leather trade with California, and Italian fashion house Versace recently announced an end to their use of kangaroo leather. Ultimately, though, these campaigns don't always have the desired effect. Skins and meat may simply go to waste when kangaroos are culled. And, if commercial hunts are no longer profitable, there is a risk that, rather than fewer kangaroos being shot, they are killed by less skilful shooters.

An extensive review of welfare codes, completed in 2020, offers more guidance to shooters. Not everyone thinks this went far enough, nor that it involved sufficient consultation. Aboriginal attitudes to kangaroos are not widely understood and they are generally excluded from decision-making. For some Aboriginal people the commercial harvest of kangaroos is culturally offensive. However, we shouldn't see Aboriginal culture as homogeneous. Culling of kangaroos without using the carcass is considered offensive to some groups, whereas others consume only the tails of kangaroos.

Once again, when we talk about wild animal welfare, it's important to consider the inconsistencies. There may be welfare problems associated with hunting kangaroos, but eating meat from wild animals is arguably better for welfare than eating meat from livestock. If we take into account the suffering that can be imposed on domestic animals, such as transportation and death in an abattoir, a commercial kangaroo hunt with skilled shooters starts to seem a reasonable alternative (as well as reducing meat consumption overall).

As a result, plenty of people in Australia and beyond seek out wild meat as a more ethical option.

Around Europe, people shoot species such as reindeer, moose, wild boar, seals and bears. Fox hunting may have been banned in Britain, but killing of game birds such as grouse and pheasants is big business, with millions of game birds being shot each year. The US Fish and Wildlife Service sees hunting as 'a wildlife management tool and an outdoor tradition', and the game meat available includes rabbit, duck, quail, venison and bison. In Japan, deer and boar are hunted for meat and to control numbers, and, controversially, whales and dolphins are hunted too.

Hunts range from unsustainable harvests of uncommon species to population control of abundant species. In some cases, such as the venison produced from Oostvaardersplassen, meat is simply a by-product of a cull. In other situations wild meat comes from species that are causing ecological damage – wild boar populations, for example, are increasing in Europe and cause problems for farmers, and to other wildlife such as snakes and lizards. The situation still isn't simple, as eating boar won't necessarily reduce the population, and there are instances of wild boar being released illegally to top up numbers for people to shoot. Elsewhere, topping up populations in this way is legal – the majority of pheasants shot in the UK have been bred and released.

Not all wild meat has the same welfare standards and environmental credentials, of course, but if you are looking to buy wild meat it is possible to make informed choices. First, you can make sure your game meat is wild rather than farmed in the same way as livestock, because farmed meat can have a much higher environmental impact. A study in Scotland showed that farmed venison had a higher carbon footprint than beef or lamb. You can also look for certifications, such as the Scottish Quality Wild Venison (SQWV) label. Butchers should be able to answer questions about where their meat came from, and may stock different wild meats in response to demand.

From an environmental perspective, another important choice is to buy local meat, as game meat is often transported bizarrely long distances. Scotland produces around 3,600 tonnes of venison each year, mostly from the wild, and about a third of this is exported. A similar amount, however, gets imported into the UK each year, including from New Zealand. In Italy, farmed venison is imported, even though locally hunted meat may go to waste.

One of my most vivid childhood memories is of a grass snake taking a frog from my garden pond. By the time we spotted this, a back leg had disappeared into the snake's mouth, and the frog's fate looked sealed. Without a pause for thought, my mother removed her sandal and waved it at the snake. This had the desired effect of prompting the snake to release the frog and retreat into the undergrowth.

Looking back, I think how privileged I was to have witnessed something amazing just a few metres from my childhood home. My instant reaction, however, was very different: the snake was stealing one of our frogs. My father built that pond and we owned the garden, so surely I had more right to the frog than the snake did? And didn't we have a responsibility to protect our frog from harm? But, from what I remember about the conversation we had afterwards, my mother realised that she had acted hastily and pointed out that the snake needed to eat. For me, this was an important step towards understanding the harshness of nature.

In the decades since that episode, my values have repeatedly shifted, as have the values of society as a whole – we're more aware of the havoc we are wreaking on wildlife, and animal welfare has moved higher up our agendas. Our concern for animal welfare focuses largely on domestic and farmed animals, as the rise in veganism shows, but it is spreading more widely. From academics arguing that we shouldn't kill in the name of conservation, to wildlife film crews intervening to help their subjects, attitudes to wild animal welfare are nuanced and changing. This chapter has provided just a snapshot of these debates.

I've also become more aware of the inconsistencies. At the time of the snake incident we had a pet cat and, although we did our best to rescue any wildlife from her jaws we generally failed. That didn't factor into my thoughts about the snake, any more than the fact that I ate meat. Now, however, as a conservationist, I'm mindful that my judgements are inconsistent. I'm critical that birds of prey are poisoned by people who consider them 'vermin' on moorland managed for game birds, yet I'm comfortable with rats and cats being poisoned on Floreana.

I stand by these values. I believe that protecting the unique fauna of the Galápagos is a more valid reason to kill an animal than for sport. However, even the notion that it's acceptable to kill rats to prevent an extinction is up for debate. And so is the language we use. In a respectful debate on animal welfare, it may be better to avoid emotive analogies, such as 'blood money' for the fees paid by trophy hunters. At the other extreme, benign euphemisms such as 'harvest' and 'take' can allow us to forget that we're talking about killing. The killing of animals is messy and complex.

CHAPTER 6

A HUMAN LANDSCAPE

Yellowhammer vs Scottish crossbill (and setting aside half the planet for nature)

'Only by committing half of the planet's surface to nature can we hope to save the immensity of life-forms that compose it.' That was E.O. Wilson's bold proposition in his 2016 book *Half-Earth*. When he published this manifesto, the 86-year-old naturalist captured imaginations with what seems a simple and elegant idea. But the proposal created a rift that conservation is struggling to navigate. Is this really the solution to halting the sixth global extinction, or the musings of someone out of touch with the reality and values of modern conservation? And what does allocating half of the Earth to nature mean? Where will these protected areas be, and what might they look like?

It's a question I find myself pondering when I visit Park Grass, the world's longest-running ecological experiment. Set up in 1856, this site – a 2.8-hectare meadow – is part of Rothamsted Research in Hertfordshire, the agricultural field station where I completed my PhD. Its original aim was to investigate how fertiliser inputs affect hay yields, but it quickly became clear that the site could tell us more and it is now being used to study the wild plants growing there.

The flowers make the hay meadow a beautiful sight. Standing next to the experiment, I am struck by the colours, the insects and the birdsong. This is land which definitely counts as the human half in E.O. Wilson's division, but try telling that to the bees. Alive with wildlife, the meadow actually carries many of the benefits that Half-Earth proponents use to justify protecting nature. Meadows store carbon and provide a home for pollinators – and, in fact, if we want pollinators to increase food production, then meadows may do more than large nature reserves. Vast protected areas that are separate from farmland (as E.O. Wilson imagines nature's half of the earth) won't contribute to pollination of our crops. This suggests that the Half-Earth Project can't be the whole answer. We also need to protect nature in human environments.

Park Grass gives a concerning insight into this – species that were present in 1865 are now missing, and even in somewhere so well studied we don't know why. I have a pertinent reminder of this in the soundtrack to my visit: '*a little bit of bread and no chee-eese*'. This is a mnemonic for the yellowhammer's song that I recall from Enid Blyton's books – Julian hears it maddeningly repeated in *Five Go Off in a Caravan* – and it is a call that has its place in grander European culture, as the inspiration for the dramatic opening bars of Beethoven's Fifth Symphony. The call rang out at Park

Grass and it once filled the British countryside, and much of Europe and Central Asia. But, as farming has become more intensive, this striking little bird has declined by 50 per cent in Britain since the mid-1980s, and populations are falling elsewhere too.

In Britain, yellowhammers have suffered from the loss of hedgerows and from pesticides reducing the numbers of weeds and insects, creating an absence of winter food. And they are by no means alone, with widespread declines in farmland birds across the country. The way forward may seem obvious: reverse our farming trends and halt the declines. But not so fast. There are trade-offs to be made, and Park Grass provides a classic example.

The Park Grass field is divided into experimental plots, which reveal that management has a huge effect on the plant diversity found in hay meadows. The difference between plots is obvious at a glance – it doesn't take a botanist to see that some host more flowers, and the most striking difference is the presence or absence of snake's head fritillaries, with their deep purple flowers drooping down like bells. This variation between plots is related to fertiliser inputs. Adding nutrients allows a few plant species to dominate, and reduces microbial activity along with the number of earthworms in the soil.

The dilemma comes because the fertilised plots – with a lower diversity of wildlife – are the more productive ones, allowing a bigger harvest of hay. This is at the heart of the tension between food production and wildlife conservation: higher crop yields generally mean less wildlife. If we reduce yields, then we need more land to grow our food, and where's that land going to come from?

Almost 40 per cent of the world's ice-free land is used for agricultural production, and if you combine that with cities and other human land uses, it's clear that we can't take more land for farming and simultaneously save half for nature. However, if we try to feed the growing human population on the land we currently use for agriculture, it will be harder to make farming wildlife-friendly. This conundrum has prompted the sharing-versus-sparing debate. In a 'sparing' scenario, intensive farming spares land for use by nature, whereas 'sharing' means farming and nature share the same area. Organic farms would be a good example of sharing – they have more wildlife than conventional farms but tend to have lower yields so need more space.

The dilemma is partly scientific: how many species will benefit from the different strategies? Early research was done on birds in Asia and Africa, and the results were remarkably consistent, showing that more bird species benefit from land sparing. The same pattern emerged for trees, butterflies and dung beetles in different parts of the world. The reason is that a greater diversity of species can live in the habitats that exist because of land sparing, whether this is woodland, wetland or moorland. For lots of species this is the only way we can help them – they rely on specific habitats, so won't benefit from wildlife-friendly farming.

The concept seems quite straightforward in tropical regions, as it is easy to imagine intensifying agriculture to save more space for rainforest. In Europe things are different; in Britain, for example, almost the entire landscape has been more modified by people, so we are not protecting ancient woodland by producing our food in a smaller area. However, the same principle seems to hold: studies of British birds show that more species benefit if we combine intensive agriculture with semi-natural habitats rather than switching to lower-intensity farming.

The winners from land sparing are habitat specialists, and British birds in this category include the Eurasian bittern, the whinchat and the Scottish crossbill. The Scottish crossbill is the only British bird to be found nowhere else in the world. It is just a little larger than the greenfinch, and the male's dusty red plumage blends into brown wings, while the female is a yellowish brown. Crossbills get their name from the shape of their beak, which ends in sharp, overlapping points. In the Scots pine forests they use this to force open pine cones and reach the seeds inside – an amazing adaptation that allows them to thrive in the forest but is of no benefit on farmland. The species is therefore absent from agricultural land, and instead has its stronghold in the Cairngorms National Park, a protected area that would come under E.O. Wilson's idea of nature's half.

Protecting the Cairngorms also benefits other forest specialists. Crested tits, ptarmigan and pine martens are just a few of the many species which thrive because this land is spared for nature. But setting conservation priorities is about more than species counts – we have some difficult value judgements to make. Do we want to walk through farmland and hear the yellowhammer's song? Or are we happy to trade farmland birds for forests and wetlands? And, if we do prioritise natural habitats, which do we choose and where?

At present, the land we've left for nature includes deserts, glaciers and mountaintops, so maybe the focus should shift to more fertile land. But this comes with challenges, and increased farming intensity won't automatically lead to land being saved for nature. It's a similar problem to energy-efficient technologies, which can lead to increased demand rather than lower consumption. We may just keep farming the same amount of land but waste more food because it's cheaper, or feed more to livestock, then eat more meat and dairy products. This is the opposite of what needs to happen – a key part of using less land for agriculture is making our diets and food systems more sustainable. If land sparing is going to work, we also need policies to ensure land is taken out of agricultural production. Knepp (see Chapter 5) is a perfect example of how this can happen, with a formerly unprofitable farm transformed to a home for rare species such as turtle doves and purple emperor butterflies. It shows promise that, in theory, farmland can fast become space for nature.

The challenge, however, isn't simply to allocate land (and sea) for nature – we need to decide what to do with it. We face complex trade-offs and competing goals, which include sequestering carbon, protecting species, alleviating poverty and managing the cultural landscapes that shape our identities. Some of these trade-offs are highlighted by the high-profile debate about rewilding. Should we abandon our desire to manage and control, and instead leave nature to find its own course?

Questions about what we want from our land are shaping the strategies for Heal, a British charity set up to encourage rewilding. Heal doesn't yet have any land to its name, but it has bold plans to buy up areas of the British lowlands. Its first employee is Hannah Needham, who is leading a founding team of volunteers. I spoke to her just after the fledgling organisation had launched, and she explained her vision was to bring nature to people, rather than people having to seek nature, with all the environmental costs of travel.

Rewilding means different things to different people, but a focus for Heal is reinstating natural processes and letting nature find its own way. Around Britain, many nature reserves are intensively managed, yet despite all this work nature continues to decline. Rewilding offers a different solution, and Hannah believes this is one reason that Heal is attracting attention: 'Letting nature steer itself is an attractive proposition, as we can unite in the idea that humans don't always know what's best. It gives us a bit of hope.'

This hasn't always been Hannah's outlook. Although as a child she loved forest in nature documentaries, she took it for granted that fields were large and square. Like the rest of us, she saw a regimented world where wildlife was tamed. That changed when she got her first traineeship in conservation and she became inspired by the idea of rewilding. However, the feeling of excitement was twinned with a feeling of helplessness. How was she going to rewild without land? So she started a village wildlife group, 'with the aim of rewilding people'.

Rewilding people is the big idea of Hannah's work at Heal, and human well-being is core to the charity's mission. Heal was launched in 2020 at a time when the hashtag #SolaceInNature was trending on social media – when the news was gloomy, during the pandemic, people increasingly turned to nature. Heal sites are therefore intended as places where people can find solace and joy, and locations will be chosen partly based on accessibility for people on lower incomes. There are countless studies that show proximity to nature, even a small patch of urban greenery, is beneficial to our mental and physical health. It pays back in economic terms, with less pressure on our health systems, fewer days off work sick; studies have also shown that hospital patients with views of nature recover quicker.

The knowledge that nature provides essential services adds a new dimension to rewilding's challenge of how much to intervene. Do we simply let nature steer its own course right from the start, or is habitat management important so that we encourage a particular outcome? For Hannah, the threat of climate change makes the argument more complex by adding a time pressure. We know that in some situations increasing forest cover can help fight climate change, but tree regeneration can be slow unless there's a seedbank in the soil. Adding those seeds may therefore be better for carbon storage than simply buying land and leaving nature to its own devices. While creating plantations with rows of identical trees can be an ineffective approach to carbon storage, planting a diversity of local species can help regenerate a forest. This is sometimes being done in creative ways. In Chile, for example, Border collies wearing backpacks full of seeds have been trained to run through areas destroyed by wildfires. As they run, they leave a trail of seeds.

The dilemma of whether rewilding should start with positive interventions brings us back to the question of exactly what we hope rewilding will achieve. As Hannah reflects: 'If what we really want is to restore carbon capture through healthy soil and tree growth, do we give nature a helping hand, and how much? If we set an end goal, what sets us apart from traditional conservation? Is that really rewilding?'

For some of us, our support for rewilding comes from a philosophical outlook on the appropriate relationship between humans and the rest of nature. We might hope that, rather than exerting our power over nature, rewilding will allow us to develop a new relationship with the natural world. The appeal of rewilding can also have a more practical element: humans don't always know how to manage nature effectively, particularly as the climate is changing. This is a guiding principle for Hannah: 'Our systems of life are so complex, and I don't believe we can predict every single behaviour and every single role of every organism. Nature will adapt, and it will adapt in ways we don't anticipate.'

Some objectives, however, can only be achieved with human interventions, in the short term at least. To enable peatlands to store carbon, for example, they need to be actively restored. Peat releases carbon as it dries out, so a first step is to re-wet the soil. This can be done through blocking drainage ditches to retain water or laying sacks over exposed areas of peat. In some cases, restoration also means removing trees that were planted in peatlands during the last century. Trees take water from the soil and their leaves intercept rainwater that would otherwise have reached the bog below, reducing the peat's capacity for storing carbon and water. This removal of trees goes against many people's intuition and against the 'leave it be' attitude of rewilding. However, such active restoration of peatlands will have local benefits such as improving water quality and reducing flood risk, as well as enhancing carbon storage. Ultimately, it will recreate an ancient cultural landscape and a greater diversity of wildlife. Management is the best way to achieve this.

The question of how much human intervention is warranted is going to be difficult for Heal to navigate, and will depend on their choice of sites and exactly what they hope to achieve. Hannah is acutely aware that letting nature follow its own path may lead to trade-offs. 'Can you imagine the dilemma if we picked a site that had a particularly rare wild flower on it? Allowing nature to do its thing could mean trees end up shading the flower out. In rewilding there has to be some acceptance of the consequences.'

Goals such as species conservation and carbon sequestration are sometimes at odds. And, given that no one patch of land can be optimised for every purpose, we need to view conservation at different scales. Part of this is understanding that the benefits of carbon storage are realised on a global scale, whereas other benefits may only be felt locally. While local conservation choices need to be driven by local people, they would ideally take into account how that area of land fits in with the broader landscape. Perhaps what we need is a mosaic of habitats being managed in different ways, and with some not being managed at all.

So, rather than a simple choice between land intensively managed for humans and land left free for 'wild nature', we might have some intermediate land. For example, small areas could be farmed at very low intensity to support wildlife such as the cirl bunting, a relative of the yellowhammer. This land is shared by humans and nature – managed for wildlife but still producing food. Likewise, wild flower meadows bring great benefits, from flood control to cultural identity, but are at risk if we dedicate large areas of land to rewilding.

Meadows are already in steep decline in Britain, with 97 per cent lost since the 1930s. Within living memory, it was possible to walk from Stratford-upon-Avon to Birmingham entirely within meadows, something unimaginable today. Meadows require ongoing management to maintain habitat for plants which don't thrive in forest. So, if giving land back to nature means rewilding, then meadows won't feature in a Half-Earth scenario. The British charity Plantlife highlights this predicament. What some people call 'giving land back to nature' is viewed by Plantlife as 'land abandonment'. They point out that meadows are rich in pollinators and other invertebrates, along with mammals and birds such as hares, voles, curlews and kestrels. For some of these species, managed habitat is better than rewilding.

Similar debates are taking place in Norway about farmland and nature. Should food production be kept at the current high levels? Or should it be returned to past, less intensive levels? Or the land allowed to be reclaimed by nature? The Norwegian mountains were once covered in birch forest, and the landscape we see today results from two centuries of grazing. The open views and rocky outcrops which many people now think of as 'wild nature' is nothing of the sort. The mountains have been kept forest free by livestock. But recently that has been changing.

Until the middle of the last century, farmers in many parts of Europe spent the summer at high altitude grazing sheep and cattle on mountain pastures and then returned to villages lower down when the weather got too harsh. However, this 'transhumance' has come to an end in Norway, as elsewhere in Europe. There were many reasons for people abandoning this way of life, including agricultural policy. The Norwegian government offered subsidies to ensure that their livestock farmers could compete in the single market, which applied only to farmers with larger herds. This prompted many smaller-scale famers to sell their livestock. The result is that sheep are still farmed, albeit in smaller numbers, but the methods of farming are very different. Rather than a low density of sheep across the landscape, there are some areas without sheep and others where they are kept at high density.

This high density has brought problems. Overgrazing releases carbon from the ground and reduces plant cover, which can increase the risk of flooding and soil erosion. There are also negative consequences for wildlife – plants are the base of mountain food webs, so the impacts ripple through to many other species. The new landscape created by high sheep densities brought benefits to farmers, but at the expense of nature and the benefits it brings the rest of us.

However, returning the mountains to forest has its own drawbacks. By preventing trees and shrubs from advancing, the sheep create an open landscape which brings cultural enjoyment to many people. Given the long history of livestock grazing, it perhaps isn't surprising that many people favour the familiar landscape created by sheep. People have come to expect a form of nature that has been modified by people and livestock, and the stunning rocky landscape with wide views uninterrupted by forest is ideal for hiking, skiing and other outdoor activities. This recent baseline for what the natural world should look like has no more moral relevance than a prehistoric baseline, yet this is a form of nature Norway now aspires to protect.

Norwegian ecologists have proposed a middle ground. Keeping sheep at a low density maintains the soil carbon stores alongside the cultural benefits of the mountain landscape. We can gain the meat, wool and income from sheep farming without problems such as soil erosion. And it is not just humans who benefit from an open landscape created by grazing. While some species declined when sheep modified the landscape, returning the mountains to birch forest doesn't necessarily protect the greatest number of species. Norway keeps a Red List of species which are threatened in the country, though some are common elsewhere in the world. It turns out that what's best for many of these species isn't landscapes untouched by humans. A high proportion of Red-Listed species in the Norwegian uplands are threatened because there are fewer sheep in the mountains. Small herbaceous plants, for example, need sheep to prevent the trees from growing and taking their light.

Elsewhere, too, leaving nature to its own devices doesn't always bring the maximum benefits. An interesting example comes from Slovenia, where political changes caused farmers to abandon agricultural land in the 1950s. Forest naturally reclaimed the land, which reduced flooding and erosion. This brought benefits, but the forest is thirsty and can cause rivers to run dry in summer. Seasonal river beds are the natural state, but people who rely on the water need rivers to flow all year round. If we want a regular water supply, we need some human management of the forest.

Studying these impacts can help clarify the effects of managing nature, but they don't provide conclusions about what we should do. Different people and wildlife benefit from different management, so inevitably there will be opposing outlooks. But it's an advance if, at least, we recognise that there's no historic baseline to aspire towards.

In Norway, people have embraced the reduced forest cover that made way for food production. Elsewhere, however, the long cultural history of forests being thinned and cleared for agriculture is viewed far more negatively. This is particularly apparent when changes to forests are made by Indigenous peoples, whose lands and lifestyles are often very far from our experience.

We like to think of Amazon rainforests as untouched by human hands, but the reality is that Indigenous societies have modified the forest through shifting agriculture for thousands of years. Sometimes known as slash-and-burn, this involves making small clearings in the forest called swiddens. Crops such as manioc and cassava are planted in these forest gardens, and they are managed to stimulate new growth and attract game for hunting. The gardens provide food for peccaries, for example, which are pig-like animals that have been managed by native people to the point that they are seen as semi-domesticated.

Native peoples of Amazonia deliberately over-plant crops to compensate for losses to animals. The welcoming of animals into their spaces is in sharp contrast with Western farming, and with ideologies that put up boundaries between humans and the rest of nature. There is also a striking difference in understandings of land ownership. Swidden agriculturalists gain ownership of their gardens by transforming patches of forest, and then low soil fertility means that they are abandoned after a few years. As the swidden becomes reclaimed by the forest, it is no longer anyone's property, leaving it free for the next settlers. Former gardens and villages are seen as potential spaces for human habitation because they bear the marks of people, and the cycle continues.

Whereas most of the world manages the natural world as resources and commodities, management by Indigenous peoples is often driven by sacred relationships. Using resources is a privilege that comes with stewardship responsibilities – it is a sacred exchange with nature, not just an extraction. In many Amazonian societies, for example, game animals are owned by spirit-masters who negotiate the supply of game with human shamans. Such sacred relationships occur in different forms around the world, with spiritual beliefs protecting nature. In India, for example, sacred trees have been tended for generations. In Bangladesh, species such as the mugger crocodile and rock pigeon are protected because of local beliefs. The Ashanti people of Ghana avoid overexploiting forests through taboos and practices such as performing rituals before certain tree species can be harvested.

Despite the fact that Indigenous peoples have been sustainably managing forests for millennia, there is a long and troubling history of disregard for Indigenous ways of life. Attempts to convert Indigenous peoples to Western ideologies still continue in the twenty-first century, and can disrupt relationships with the forest. An ongoing struggle is apparent in the situation of the Trio, who live on both sides of the border between Brazil and Suriname. The Trio are known by a variety of names, and refer to themselves as Tarëno ('the people here'). They live in remote locations that are far from any roads, and hard to access by river due to shallow water and dangerous rapids. They live primarily from swidden agriculture, with bitter manioc being their main crop, as well as hunting and fishing.

Their semi-nomadic lifestyle was initially disrupted by the arrival of Christian missionaries, whose actions supported the government's desire to control the country's borders and 'civilise' the people. The first missions among the Trio of Suriname began in 1960, following a military programme to cut airstrips in the forest, known as Operation Grasshopper. One of the first changes the missionaries made was to concentrate Indigenous peoples into villages, and transform them into sedentary farmers. Before this, village populations rarely grew to more than about thirty people, and settlements regularly shifted. Now, the 2,800 Trio are composed of distinct groups which came together in larger settlements around mission stations.

The transition away from traditional subsistence living prompted by the arrival of missionaries has had effects on the forest. Traditionally, there is a strong stigma amongt the Trio against buying and selling meat. This would conflict with the culture of sharing, and threaten the joint identity it creates. The low human population in southern Suriname means there's no need to overhunt in a sharing economy. However, now that the Trio are concentrated around mission stations, there is access to markets, and bush meat is sold to Paramaribo, Suriname's capital city. Other forest products are traded too, including animals for pets, so the Trio are increasingly part of the cash economy.

Such transitions have been made easier by conversion to Christianity, with all animals seen as under God's dominion, which makes hunting them legitimate. Some missionaries have promoted this outlook to extreme levels. One even shot all the caimans he could find, knowing they were of mythical importance to the Trio. Inevitably, much of the meat went to waste. With Trio living at much higher densities and hunting meat to sell in the city, the previous barriers to overhunting have been lost.

The Trio's knowledge of the forest has inevitably declined due to relocations, the introduction of diseases and the outlawing of local traditions and language. Without access to the forest, elders are unable to transmit traditional knowledge and cultural practices to future generations. Food sharing and gathering food together are important for social bonds and passing on knowledge, yet these cultures are being eroded. This is compounded by the fact that traditional knowledge developed over millennia may no longer apply because humans have changed the ecosystem beyond recognition. Once people have been alienated from their ancestral lands, it is easier to finally remove them.

Indigenous lands in Suriname are also under threat from mining and the timber industry. With no formal recognition of Indigenous land rights, the government is free to allocate Indigenous lands to mining, oil and natural gas extraction. Mining for gold and other minerals is the leading cause of deforestation, threatening Suriname's exceptionally high level of forest cover. Forest covers over 90 per cent of Suriname, and it is home to a huge diversity of plants and animals, including endangered species such as the jaguar, harpy eagle and giant river otter. It's a loss that should concern us all. These forests are not only sustaining the animals and people that live within them, they're providing a stable supply of fresh water and contributing to climate regulation that brings global benefits.

With Indigenous lands representing about a quarter of Earth's land surface, there is a huge potential for synergy between wildlife conservation and upholding Indigenous rights. However, the reality is that Indigenous peoples' rights to the forest has sometimes been denied not just by cultural subversion or a drive for power and profit, but by conservation. National parks and protected areas have a long history of displacing Indigenous people, removing them from the land that sustained them and ending their way of life (and, in extreme cases, ending their lives). Even when people are allowed to remain on the land, there can be limits to resource use – they cannot hunt or collect food, medicinal plants and firewood on their ancestral lands. In the past, protected areas have been created without meaningful involvement of those most affected, to the point that sometimes people aren't even aware that they're now living within a protected area.

However, this colonial approach to conservation is being challenged, as we will see in the next chapter. The preservationist outlook that separates humans from nature faces a backlash, and local people are increasingly recognised as forest guardians. Of course, we can't simply turn back the clock – Indigenous lifestyles may no longer be sustainable now that populations have changed, knowledge has been lost and forest has been destroyed. But, while the future won't look like the past, there are signs it can be bright.

When local knowledge is combined with Western sustainability science, it is possible to find new ways of living sustainably in the forest. An example comes from the Asháninka tribe in Peru, whose slash-and-burn agriculture had become unsustainable. The rate at which they were clearing forest to grow crops such as cassava, cocoa and banana became very high when the population began clustering around new roads, leading to worrying deforestation. The British charity Cool Earth has worked in partnership with Asháninka people since 2008 to facilitate a shift to sustainable livelihoods. They provide grants so that the communities have no need to sell trees to loggers. And they have worked with local people to improved farming practices, ensuring that each patch of land can be used for longer, which reduces the need for forest clearance.

This is just one example of how bringing excluded people into negotiations as partners has great promise for nature conservation. In Bolivia, Brazil, Colombia and Peru, granting tenure to Indigenous lands has proved an effective way to reduce deforestation. However, we are a long way from securing the protection that Indigenous peoples need. Including them as true partners in forest conservation is vital, but also complex, partly because they are generally invited into discussions run on terms dictated by outsiders. Land rights, for example, can be debated in ways that are alien to them – our idea of private land ownership often fails to resonate with communities who hold their territories and resources collectively.

The Trio have not yet found the protection they need. The Surinamese government doesn't recognise Indigenous forms of land ownership, and swidden horticulture is seen as forest degradation. However, even though Suriname is the only South American country with no official recognition of land rights, Indigenous peoples there have made great progress towards forest protection. After an extensive engagement process with communities, Indigenous leaders in south Suriname signed an Indigenous Declaration for the protection of 7.2 million hectares of tropical forest in 2015. This area is about 40 per cent of Suriname's land surface and almost the size of Scotland. By signing the document, the Indigenous communities declared the creation of the Southern Suriname Conservation Corridor, which was later renamed Trio and Wayana Protect Land and Nature in Southern Suriname (TWTIS) so they can identify themselves more with the project. Sadly, though, TWTIS hasn't yet gained legal recognition as a protected area. For this to happen, Surinamese nature protection laws need to be updated to allow co-management by Indigenous communities and give scope for them to lead sustainable livelihoods within protected areas. New legislation was submitted to Parliament in August 2018, but as of 2021 it has not yet been approved. Until it is, the TWTIS cannot be protected by law.

Throughout this chapter we've encountered complex trade-offs in deciding how to manage land for food production and for wildlife, and whether to manage it at all. Scientific knowledge is essential to making these decisions wisely, and more research is needed to explore questions such as whether we can maintain soil fertility under intensive agriculture, or how can we increase yields without causing declines in wildlife.

We also need better data about the impact of wildlife on farming. Can the species which thrive thanks to rewilding bring benefits for food production? One issue is that, although farming relies on wild species such as pollinators, these tend to be habitat generalists that are common in human landscapes. In contrast, most of the species in large wildlife areas don't contribute to agriculture. Instead, most agricultural benefits come from a small number of species, and they often don't travel far. The best way to ensure these generalist species support agriculture can be to set aside much smaller patches of land for nature, such as wild-flower borders adjacent to farmland. Having patches of wildlife habitats within the farmed landscape can bring great benefits, including reduced erosion, providing a source of pest controllers such as wasps, slowing the evolution of pests and weeds, and controlling floods. Even very small areas can bring benefits – buffer strips of vegetation around farmland, for example, can help reduce the runoff of chemicals and sediments, increasing water quality. Trees can provide windbreaks to protect soils, livestock and crops.

Sometimes the benefits will come immediately – restoring forest fragments contributes to higher oilseed rape productivity in Brazil, for example. In other instances, we could produce more food in the short term by farming these patches rather than leaving them for nature. However, we need to consider long-term food security. High levels of production can't always be maintained, and agriculture which relies on the services of wildlife rather than artificial inputs may be more sustainable. This is a key point, because intensive agriculture will only spare land for nature if it can be sustained in the long term.

We must also consider the long-term benefits of larger areas set aside for nature, something we will explore in the following chapter. These large areas may, for example, ensure the survival of species that will play important roles in the future. Increased knowledge helps us make sound decisions. However, the science isn't enough. There are also value judgements to be made, both about the species we want to prioritise and the way we want to live our lives.

Our lifestyles feed in to many of the trade-offs that the rewilding and Half-Earth debates are grappling with. We can live in dense cities to spare more land for nature, or spread out our settlements and make more space for nature within them. Dense human settlements might help realise the Half-Earth vision, but is that how we want to live? We have all taken joy from the nature around us, and research shows that urban nature is good for mental health, as well as bringing benefits such as air purification. Is contact with nature something we are willing to forgo so that more of the Earth can be in a wilder state?

We also face trade-offs at a much smaller scale. Even the smallest management decisions will affect people differently, and can have unintended consequences. Could reintroducing beavers cause a neighbouring farm to flood, for example? When Heal acquires its land, it will tackle such questions, as far as possible, but discussions with different people start now. Hannah's focus is consultation with Heal Future, which was set up to give young people a voice. But she is equally keen to engage with people who have different life stories – she values a diversity of people in the same way that she values a diversity of wildlife. If Heal wants to make its sites accessible, they need to hear from people with disabilities and people from varying backgrounds, including those who don't normally spend time with wildlife. Only then can they fulfil their desire to close the rift that has opened between humans and the rest of nature.

CHAPTER 7

FORTRESS CONSERVATION

Oceanic whitetip sharks vs. Whitetip reef sharks (and the virtues of protected areas)

In 2001, a single fishing vessel travelled to the Phoenix Islands – a Pacific archipelago north of Samoa – in search of sharks. It found them in abundance and the crew fished in the waters surrounding four of the islands. The effect was immediate and dramatic: shark numbers in the area dropped by 90 per cent. The population slowly recovered but four years later a second shark-finning vessel appeared, and again began longlining. Photos show rows of abandoned shark carcasses filling the beach, bloodied where their fins have been removed. Once again, the effects were devastating. Just one fishing vessel had wiped out almost a whole population.

Thankfully, there's no danger of a repeat today. Since 2015, shark finning has been banned in these island waters, along with all commercial fishing, after they were declared a no-take marine protected area (MPA). The Phoenix Islands Protected Area (PIPA) is, in fact, one of the largest marine protected areas in the world, preserving both shallow coral reefs and deep ocean.

The islands and ocean protected by PIPA are part of the Republic of Kiribati, whose waters span the equator. The Phoenix Islands are the centre of its three groups of islands, formed by long-extinct volcanoes and now capped by coral atolls. With golden beaches, palm trees and clear, shallow waters, they are the epitome of a tropical paradise, and the same beauty can be found below the water, with extensive reefs in almost perfect condition. The remoteness and lack of freshwater on the islands means the various attempts at colonisation never lasted long, and so the surrounding seas have been relatively untouched. As a result, many endangered and endemic species are found there, along with migratory birds, mammals and sea turtles. Over 120 species of coral and 500 fish species have been recorded, including thriving populations of parrotfishes and wrasses overfished elsewhere.

In many ways, the story of the Phoenix Islands is a triumph. This tiny island nation offers a high level of protection to around 11.3 per cent of its ocean. And this protection is paying off – populations of previously exploited species, such as giant clam and coconut crab, have been recovering since their MPA designation. However, the story also illustrates some of the complexities and trade-offs involved in protecting the oceans. MPAs are rapidly increasing across our planet but not everyone is convinced that they are the solution. In theory, they increase fisheries catches while simultaneously preventing destruction of habitats. But not everyone sees them as win–win scenario; some view them as 'ocean grabs' in which powerful conservation groups exclude local people from the waters that nourish them.

I discussed this issue with Dr David Obura, a Kenyan scientist working to create sustainable marine environments. His involvement in PIPA goes back to the early 2000s, when the idea for an MPA was just developing, and he has worked extensively with marine protection in Africa and beyond. He explains his outlook: 'Protected areas are the core zones of a much wider space that species use – they're a bank that sustains everything else. We therefore need to keep them in as good a condition as possible, but they can't be our only focus. The whole ocean needs to be managed in a fair and sustainable way.'

When David first dived around the Phoenix Islands. he saw many species that would benefit from an MPA, including three species of reef shark: the grey, the whitetip and the blacktip. Still today, a dive on the Phoenix Islands' corals reefs is likely to reveal sharks sleeping in caves under reef overhangs. The three reef species are easy to distinguish based on the colour of their fins; the whitetip reef shark has a bright white cap to its dorsal fin. It is a small, slender species, growing to about 1.6 metres long on a diet of fish, crustaceans and octopus. Like many sharks, its population is decreasing globally, and the International Union for the Conservation of Nature (IUCN) classifies the species as vulnerable.

The whitetip reef shark is found on coral reefs in large parts of the Pacific and Indian Oceans, and throughout most of this range it is under threat from fishing. Sometimes it is a target of fisheries, but often it is caught as a bycatch and retained for its fins, meat and leather. PIPA's protection is vital because the whitetip reef shark and its relatives have very specific habitat requirements and spend much of their time in small home ranges. The protected reefs provide them with shelter from predators (including people), plenty of food and perfect conditions for breeding. Nikumaroro island, for example, circles a shallow lagoon – cut off from the wider ocean at each low tide – that is an ideal nursery area.

The Phoenix Island whitetip reef sharks demonstrate the value of MPAs for conservation. However, not all sharks can be saved by protecting these relatively small areas. Oceanic whitetip sharks are much larger and stockier than the whitetip reef, with broad, paddle-shaped fins, and as the name implies they roam over large areas of ocean. They used to be a common sight in both tropical and temperate seas worldwide, but are now classed as critically endangered. Like the reef sharks, a major factor in their ongoing declines is fishing – they are caught for their fins and meat, and as accidental bycatch. Their inquisitive nature makes them easy to catch, and long generation times mean the population is slow to recover.

The oceanic whitetip shark's lifestyle of roaming the ocean means that protections need to be in place over far wider areas than even the largest MPAs. They already have some protections, notably from CITES (Convention on International Trade in Endangered Species of Wild Fauna and Flora). The international trade in wildlife products is worth billions of dollars each year, with hundreds of millions of specimens traded dead or alive, and CITES seeks to ensure that wild species don't have their survival threatened by this trade. The oceanic whitetip shark is one of over 37,000 species subject to CITES trade restrictions.

Regulations are also in force to control shark fishing. More than twenty countries have banned shark-finning in their territorial waters and others have partial bans. This is huge progress, but the restrictions are by no means universal and not always enforced. Illegal shark-finning is also a concern. For the legal trade in shark fins, a shark must be brought back to port whole and used for meat (and other products such as leather). In illegal shark-finning, the fins are often removed at sea, even while the shark is still alive, and the rest of the carcass thrown back into the sea. If we're going to stop the decline in sharks, both the illegal and legal trade in shark fins needs to be addressed, as well as bycatch of sharks in nets and on longline hooks.

The lifestyle of the oceanic whitetip shark means that MPAs can't be the only answer to marine conservation – we need to care for the whole ocean. We also need to consider smaller, less charismatic species that don't receive international attention. In Chapter 2, we met some of the incredible organisms of the seabed, and these will never be named on international agreements or be part of fishing quotas. Some may be protected by MPAs, but species which occur outside these areas need other means of conservation. For example, their survival may be incompatible with destructive fishing methods such as bottom trawling.

We can't turn the ocean into a giant protected area, any more than we can enforce strict protections everywhere on land. We're therefore faced with the question of whether conservation is best served by focusing on defined areas with strict protection or on lighter protection over the whole seascape. Once again, there are challenging trade-offs and the question is made more complex as the level of protection varies between MPAs. Fishing is a particular source of tension and many people are surprised to hear the extent to which fishing is still allowed in MPAs. The UK, for example, has an extensive MPA network and yet commercial trawling is permitted in large areas. Likewise, when PIPA was founded in 2008, fishing was only banned in a very small area. As David Obura explains: 'It was a big move to extend the boundaries of protection from just the islands and reefs to include the deeper waters where tuna is caught. Initially we didn't broach this idea, because it would have been too sensitive.'

Even after PIPA became a 'no-take' zone with the highest level of restrictions, a small amount of fishing was still permitted. Commercial fishing was banned, but subsistence fishing was allowed because of the importance of seafood for island residents. This only happens on a very small scale in the Phoenix Islands, which have no permanent inhabitants (the Canton atoll is home to just forty government employees and their families). Elsewhere, though, subsistence fishing can be significant. Small-scale fishing is a huge part of coastal economies in the African countries where David focusses his work. As he says: 'Many of the communities have high levels of poverty – for many, fishing has been in their families for generations; for others, it's really an activity of last resort. It provides protein for the family and also a basic cash income that supports education and health.'

Residents of Kiribati's other two archipelagos rely on income from fishing, and the Republic receives much of its income by selling tuna fishing licenses to foreign nations, including the USA, Japan, South Korea, China and European nations. Fishing license revenue varies significantly, but for the decade from 2006 to 2015 it contributed almost 40 per cent of Kiribati's GDP. Some of this used to come from waters around the Phoenix Islands, which might seem like a powerful motivator to keep all its waters open to fishing. However, closing an area doesn't necessarily mean that fewer fish are caught overall. In fact, given that fish populations have been seriously depleted, closing areas to fishing may even be a way to increase their numbers.

Declining catches are hitting subsistence and commercial fishers alike, and some scientists believe that over a third of fish stocks now yield only 10 per cent of their historic catches. Multiple fish species, particularly those found in the deep ocean, have been fished to the point of population collapse, prompting the fishing industry to shift its targets to different species. Tuna populations around Kiribati certainly need careful management given that a commercial fishing vessel can catch an average of 32 tonnes of tuna each day. MPAs can have an integral role to play in this.

One of the arguments for PIPA being upgraded to a no-take MPA in 2015 was evidence that tuna spawned there. Tuna are migratory, so the fish that are spawned in protected waters will leave these areas as adults and become available to catch. As David says: 'If you allow them to reproduce while they are there, you may get more fish elsewhere. Tuna are protected at a specific part of their life cycle and that's to the benefit of everybody in the system.'

The positive effect that MPAs can have on fish populations outside their boundaries is known as spillover, and many examples have been reported. Researchers at University of California, Santa Barbara, concluded that we could increase global catch by at least 20 per cent if we strategically expand MPAs, which means that protecting an extra 5 per cent of the ocean could generate up 12 million tonnes more food each year. However, evidence for spillover is hotly debated and the effect is hard to prove. It is much less apparent in cooler waters, and the long-term studies needed to monitor it are difficult and expensive. It can also be hard to tell whether high fish numbers in MPAs are due to increasing populations, or because fish are attracted from elsewhere.

The impact of MPAs on fish populations is also variable. Studies show that, on average, MPAs have a positive impact on fish biomass, but this is not universal. Many MPAs are poorly enforced 'paper parks' with restrictions in place but nobody enforcing them. This is so pervasive that it has led to great scepticism about the power of MPAs to protect marine life. MPAs can give a false sense of security that the ocean is being protected, so our celebration of expanding ocean protection could be a dangerous distraction.

In the case of PIPA, reports were very positive. The Kiribati government undertook annual patrol visits, and Australia, New Zealand and France provided aerial surveillance. Kiribati enforcement officers were also present on US Navy and Coast Guard vessels operating in the area. Efforts seemed to be paying off, and Automatic Identification System (AIS) data indicate that virtually all fishing activity has indeed stopped. This kind of automatic monitoring brings the hope that protections can be enforced elsewhere, and modern enforcement technology can be amazingly effective if it is properly funded. Remote satellite advances, for example, are increasing the identification of illegal activity.

Such technologies can bring great benefits if they can keep boats out of areas where fishing is banned, but they don't necessarily mean the problem is solved. Boats may simply move elsewhere to fish, and this can increase the damage to marine life if they move to areas that are less efficient. If a boat has to cover larger areas to catch the same amount of fish, it can take more bycatch, destroy more habitat, burn more fuel and create more noise. Some conservationists are therefore calling for a move away from MPAs to a more coordinated protection of the whole ocean. Wider catch limits could ensure that fish are only caught at sustainable levels throughout the ocean. However, sustainable management has to be based on far more than quotas of commercially important species. Much of the damage is incidental, such as bottom trawls damaging the seabed, so other restrictions may need to include types of fishing gear. In David's opinion, we need both approaches: 'Protected areas are the core we need to focus on, but we also have to make sure that in other areas we're not just allowing any old activities to happen that might be very degrading.'

And the issue goes beyond the oceans. Fishing takes place in order to supply people with a source of protein. So, if fewer fish are caught, there can be negative conservation consequences on land – we will need to get more of our protein from agriculture which takes up land that could otherwise be wildlife habitat. Once again, distant conservation choices are intimately connected to our lifestyles.

Another major concern is that MPAs can't tackle threats that come from outside their boundaries. You can ban boats from discarding plastic in an MPA, but you can't ban plastic from entering its waters. And MPAs are as vulnerable as anywhere else to climate change. Protecting a single patch of ocean won't prevent it from warming and acidifying, which may have devastating effects. A protected area that was carefully planned to benefit certain species or habitats may no longer be able to support them. But, as David stresses, we need to be open to change: 'What's in the protected areas will change over time, and of course you will be losing some species, but the areas may be even more important for what moves in, and for what they can become in a warmer world.'

The need to withstand climate change formed a big part of the original argument for setting up PIPA. There is hope that it can serve as a climate refuge, and that the corals will be sheltered from the worst impacts of climate change because they're part of large reefs with few other pressures. This has been repeatedly put to the test in Kiribati, with coral bleaching causing particular concern in 2002, 2010 and 2015. Bleaching occurs when coral polyps expel their symbiotic algae, which normally provide up to 90 per cent of their energy. If recovery isn't quick, then the coral will starve. So far, the signs are that Kiribati's reefs recover well from climate stress, but we can't assume this will continue. In 2019, and again in 2020, the ocean reached its hottest levels in recorded history – and the rate of warming is also faster than at any point in the last two millennia. If the trend continues, we can't expect corals to keep recovering.

Climate change may also mean corals being affected by sea level rise, with shoreline erosion causing sediments to settle on reefs, and the potential impact of this prompted Kiribati's president to highlight the problem to the United Nations, calling for a 'brutally honest' discussion. The creation of PIPA seemed to help strengthen the islands' case.

However, the story took an unexpected turn in November 2021, when the Kiribati government announced plans to allow tuna fishing in the no-take zone once again. They stated that PIPA's endowment fund had not been able to replace fishing revenue as had been hoped, and claimed that millions of dollars of revenue had been lost since 2015. They therefore plan to shift to managing use of marine resources, taking into account that Kiribati's efforts to protect biodiversity need to 'strike a balance to enable actions that also uplift the livelihoods of our people.'

It's not clear how this situation will play out. The plans won't become official until they have been formally approved by Kiribati's parliament. Assuming the no-take rules do get lifted, it will take years to discover whether fishing is done at sustainable levels. The twist is a stark reminder that, however well protected an area is, commitments can suddenly get reversed. Protected areas often remain contentious even after they have been designated.

A striking example is the Chagos Islands, a group of seven atolls in the middle of the Indian Ocean, with a rich marine ecosystem and a troubled colonial history. Long uninhabited, they were claimed in the eighteenth century by the French, who transported enslaved Africans to work on coconut plantations. Their descendants remained on the islands after control had passed to the British in 1816 and slavery abolished. In the 1960s and 1970s, however, the British government leased Diego Garcia, the largest island, to the USA for a military base, and resettled the Chagossians in Mauritius and the Seychelles. The base is the primary reason that Britain retains its claim over the atolls, which are officially designated the British Indian Ocean Territory. This claim is at odds with the UN's International Court of Justice, which ruled the Chagos islands as part of Mauritius. Britain, however, rejected the ruling.

In 2010, the British established a strict no-take MPA around the Chagos Islands, covering an area about twice the size of the UK. This decision was guided by marine biologists, but involved no consultation with Mauritius, who maintain that it violates their rights to use the resources in the Chagos waters – and to be consulted. In 2015, Mauritius and the UK went head-to-head in the Permanent Court of Arbitration in The Hague, and it was declared that, by establishing the MPA, the UK had breached its obligations under the UN Convention on the Law of the Sea (UNCLOS). The MPA was declared illegal. The UK, however, was unperturbed, and still considers the MPA to be in place.

The Chagos MPA and lack of consultation also caused anger in the Chagossian community, who are now divided between Mauritius, the Seychelles and the UK. Like so many people who are forcibly relocated, Chagossians faced huge difficulties in establishing their new lives, and many are now fighting to return home. However, the UK government rejects their right to resettle on the islands, citing military security, the cost of resettlement and the need to protect the marine ecosystem. Some Chagossians therefore believe that the UK declared an MPA primarily to prevent them from returning to their homeland. The naval base itself is excluded from the MPA, meaning that coral reefs can be destroyed to make way for new infrastructure.

The UK government has promised to hand control of the Chagos back to Mauritius once the islands are no longer needed for defence purposes. But the lease to the US military lasts until 2036, so these promises are somewhat empty. And, even if the day comes, it's unclear what fate awaits the Chagossians and the wildlife.

The Chagos MPA is an extreme example of a colonial attitude to conservation. But in Kiribati, too, there were criticisms of how the PIPA was created through a partnership between the government and international charities (Conservation International and the New England Aquarium). Decisions about restrictions were thus driven by foreign environmental organisations, rather than consultations with local people.

MPA negotiations in Kiribati were much easier than in the African communities that David works with: there were far fewer people involved, they have the ocean at the core of their culture and they understand the vulnerability and limits of their islands. As David says, 'It was straightforward to talk to the leaders, for everybody from the top to the bottom to agree on the principles and why this is needed. It's much harder in Africa, where you have many more cultures and very different levels of need and understanding.'

Conservation groups, too, often have very different world views. Some oppose segregating humans and nature and contend that too much emphasis is placed on excluding people from MPAs; they may support fisheries management rather than bans. Other organisations have a preservationist outlook, common in the past, that seeks to protect land and ocean from any human impact. This has become known as fortress conservation, and people with this perspective on the oceans are likely to argue that strict no-take protected areas are essential for the recovery of fish stocks.

Although there is much merit to arguments for managing fish stocks through MPAs, the supporting ideology can have a darker side. The 'preservationist' heritage has led to fear and resentment towards conservation organisations, who are often seen as prioritising fish and reefs over people. MPAs can be viewed as conservation organisations claiming areas of ocean for their own agendas. However, attitudes are changing among conservationists and the communities they affect. Many environmental organisations now invest properly in local relations, showing they are in for the long term and value the interests of people, not just the ecosystem. When charities ensure their work is acceptable to local communities, they can be very welcome when people have noticed a decline in their fishing. 'In the past, people have been powerless to do anything about fish declines because everyone needs to eat and feed their families,' David says. 'Now environmental charities are key partners for local fishing communities, helping them to establish their rights and prevent degradation by external fishing fleets.'

By engaging with local communities, conservation organisations also increase the chance of success – people are far more likely to follow the rules in a protected area if they feel it has been fairly set up and managed. In particular, local management may be more respected than outside powers: David has found that many African governments have realised they don't have the capacity to centrally manage the huge number of fishing communities and so have devolved powers. 'All communities had Indigenous or local ways to organise themselves, and an elder would be in charge. They now help decide who fishes in an area, and which areas are closed for fishing, permanently or seasonally. It's really about co-management between these local groups and the fisheries departments.'

The importance of local partnerships has been brought to the forefront by the UN's Intergovernmental Science Policy Platform on Biodiversity and Ecosystem Services (IPBES). David is one of the experts helping IPBES collect vast quantities of scientific data about the natural world, and this is integrated with Indigenous knowledge. IPBES allows submissions in local languages and welcome information in different formats, including songs, videos and artworks. At its best, this kind of work goes beyond consultation with local people and treats them as equal partners. As we've seen throughout this book, many assumptions about how the natural world 'ought' to be are nothing more than cultural outlooks, and here is a chance to move beyond them. A recognition of other values and knowledge systems goes a long way towards breaking down the colonial outlooks that conservation has been guilty of in the past.

We can also point to practical success stories on land and sea, and it's now clear that community management of Indigenous lands can help combat climate change and reduce environmental degradation. A recent review of 160 protected land areas found that allowing sustainable use of resources led to better conservation outcomes, and the World Bank's Independent Evaluation Group concluded that community-managed forests are more effective at reducing deforestation than strict protected areas. This shouldn't surprise us, given that Indigenous communities had been sustainably managing ecosystems for many centuries before the concept of conservation was even imagined.

Debates about protected areas on land have some striking similarities and differences to those over marine protected areas. The central dilemma for protection of marine wildlife is that many species, like oceanic whitetip sharks, move over vast areas. But this is less common on land, where even long-distance migrants tend to have quite narrow breeding and wintering grounds, even if they are located far apart. It means that, on land, we can say with relative certainty that large areas with strict protection will protect the greatest number of species. E.O. Wilson's desire to set aside half of the Earth for nature was primarily motivated by the fact that large protected areas house more species. He is keen for nature's half to include the most diverse regions, hopefully preventing the greatest number of extinctions. This isn't a romantic idea of wilderness, but a scientifically robust way to protect most species.

Inevitably, though, it has profound implications for people who live on land that becomes protected. A seemingly innocent objective of protecting land for nature can become a form of eco-colonialism, benefitting the powerful few who have the luxury of valuing nature in a wild state. However, protected areas also have the potential to promote ecological justice, benefitting everyone, both rich and poor. The story of Half-Earth's flagship project, Mozambique's Gorongosa National Park, reveals elements of both.

Gorongosa lies at the southern tip of the African Rift Valley, midway between the border with Zimbabwe to the west and the Indian Ocean to the east. The landscape is dominated by Mount Gorongosa, with its granite peak rising to 1,863 metres. Surrounding the mountains are a rich variety of habitats, from rainforest to open grassland. E.O. Wilson first visited in 2011, although he had already been writing about its amazing diversity. Given his motivation to protect as many species as possible, it is little wonder that he was inspired. The park boasts over 6,000 known species, with many more yet to be discovered. Over 800 species of plant have been identified in its forests, which are home to endangered animals ranging from a giant fruit bat to the Gorongosa pygmy chameleon. The endemic species found nowhere else include a katydid, a millipede, a butterfly and a lizard.

In the local Mwani language, *Gorongosa* means 'place of danger', and in the last century that has been true for both people and animals. The park was established in 1921 as a colonial hunting reserve and was accompanied by removal of local people. This continued over following decades as the reserve gradually swallowed up more land, evicting residents as it grew. Big game still thrived, despite the pressures from hunting, and when it was declared a national park in 1960, Gorongosa was home to 14,000 African buffalo, 2,200 elephants and 200 lions, as well as iconic African animals such as hippos, zebras, wildebeest and a variety of antelopes.

Gorongosa's reinvention as a national park in the 1960s meant the end of game hunting. Income now came from tourism, and a new camp was built, with swimming pools and a gift shop. The Park's main draw, however, was its huge herds of iconic animals roaming the plains and forests, along with an abandoned park headquarters known as the 'lion house'. The deserted building was missing its windows, but a spiral staircase still remained on the outside, and provided the perfect way for lions to reach the flat roof. Photographs from the time show lions soaking up the Sun from this convenient lookout. Gorongosa prospered as a high-class holiday destination, attracting Hollywood stars such as John Wayne, Joan Crawford and Tippi Hedren. When Apollo 16 astronaut Charles Duke visited in 1971, he reportedly said it was as thrilling as landing on the moon.

Local people, however, were kept out. With the exception of workers, black Mozambicans were only allowed into the park by special invitation. However, the world was changing, and Europe's colonies were demanding their independence. Portugal was one of the last countries to relinquish its African territories, but in 1975, after ten years of war, Mozambique finally negotiated its independence.

Sadly, Mozambique's independence was followed by a fifteen-year civil war in which hundreds of thousands of people lost their lives. Although Gorongosa had been largely unscathed by the war of independence, the area became host to frequent battles between the RENAMO rebel army and the ruling FRELIMO government forces. The park became a place of danger for both local people and the animals it had once protected. Soldiers from both factions slaughtered the wildlife for meat and elephants were killed for ivory to trade for weapons. Hunting continued even when a ceasefire was declared in 1992. Local people, struggling to feed themselves with subsistence agriculture, set traps for whatever animals remained, and cleared ever more forest for agriculture. There was no sign of this stopping – farming practices were unsustainable, so when soils lost their fertility farmers would clear more forest, until gradually farms moved on to the slopes of Mount Gorongosa. The population of large mammals crashed to less than 5 per cent of its pre-war levels, with many species on the verge of being lost from the park forever.

Peace, however, gradually brought opportunities. The European Union and the African Bank for Development supported restoration in Gorongosa, clarifying the Park's boundaries and rebuilding some of the roads. The rapid declines in animal populations were halted, though they were slow to increase until 2004, when Mozambique's Ministry of Tourism signed an agreement with an American billionaire philanthropist called Greg Carr. This marked a dramatic turning point.

Carr had been inspired by the writings of E.O. Wilson, and recognised how special the Gorongosa Park was, even though animals were now scarce. That changed with his investment. Wildlife recovery accelerated and herbivores such as water-buck, reedbuck, warthog and impala bounced back of their own accord. For some species, progress remained slow. Where there had been thousands of buffalo, now there were only fifty. Wildebeest weren't found at all. These species were going to need more help, so in 2006 a huge wildlife sanctuary was fenced off to exclude predators and poachers. This 62-square-kilometre safe zone housed animals for release, and has helped swell the populations of buffaloes, hippos, zebras and wildebeest. The lion population has returned, too, and in 2007 lions were photographed at the lion house for the first time since the 1960s. The park was regaining some of what had been lost. Perhaps its most exciting reintroduction has been wild dogs, known locally as 'painted wolves'.

These comebacks aren't just triumphs in their own right. The animals enhance the ecosystem by playing roles such as population regulation and seed dispersal, and even those people relegated to the fringes of the park see benefits. Floodplain drainage, for example, has been improved by hippos keeping channels open. Herbivores trample and eat vegetation, which can prevent tall grasses from becoming a fire risk. A stark reminder of the park's importance came in 2009, when Cyclone Idai hit the area, destroying forest, homes and crops. The park's vegetation and soil absorbed enormous amounts of water, releasing it slowly and so reducing the peak of the flooding.

Park rangers were leaders in the cyclone relief effort, reflecting Greg Carr's vision of a park which supports human development. Local people may have been removed from the park itself, but it is surrounded by a buffer zone of almost 200,000 people. Greg's work includes a range of humanitarian projects there − schools have been built, farmers have been equipped with skills and training, and thousands of children have been vaccinated.

If we are going to set aside land for nature, this focus on the local community is essential − the people who used to rely on resources in the Gorongosa Park need alternative ways of living. In particular, equipping farmers with the skills and tools to farm sustainably in the buffer zone will mean they can use that land continuously, and not have to move into the park when their soil loses its fertility. If people can't live sustainably outside protected areas, there is no way to afford the luxury of leaving land undeveloped.

Gorongosa's reliance on the support of the local community is even more vital because there is no boundary fence to divide the people outside the park from the animals within. This comes with problems for people as well as for wildlife. Elephants are attracted to fields of crops, where they can destroy a family's livelihood in a single night. Seeing animals around your house and fields is a very different experience to seeing them on safari, and the park is working hard to help. It has had particular success with beehive fences. This involves placing beehives around fields, at elephant crossing points; crop raiding is dramatically reduced, and the honey can be harvested too.

Alongside stories of harmony between the park and the local community, however, there are ongoing struggles. Ever since Gorongosa became a colonial hunting reserve, local people have had their access restricted. The Portuguese transformed farmers into squatters and redefined traditional hunting as poaching. None of the restrictions were enforced during the war and people again came to rely on the park for survival. As well as food, they collected thatching and fire-wood, with wood and charcoal providing important income to the poorest families. The park's new managers were faced with uncomfortable trade-offs. Should they restrict human activity to sustainable levels, or remove people to priori-tise the aesthetic quality of the landscape? Greg may have bold plans for increasing the well-being of people around the park, but they needed to feed their families now and restricting access to the park impairs their ability to do so.

Under previous management regimes people would simply have been evicted. Instead, they are being encouraged to settle on agricultural land in the buffer zone, negotiating deals with communities. Park employees describe these relocations with language such as 'restoring the integrity of the park', which raises interesting questions. Ecological integrity is hard to define but tends to mean ecosystems have their natural components intact, including living organisms and processes such as fire and flooding. It also suggests that removing humans enhances the ecosystem's purity, something that history reminds us to question. Humans have been present in Gorongosa for hundreds of thousands of years, and the Stone Age tools discovered there suggest humankind has continuously altered the habitat. We shaped the landscape, and the landscape shaped us. The long history of changing human impacts doesn't tell us what kind of human activities should take place in a protected area now. We're back to the questions of what we want from a national park and who will benefit.

One new dimension to human activity in Gorongosa Park is coffee growing. There's no tradition of growing coffee in Mozambique, but it now allows farmers to make money while helping restore the forest. Coffee is planted in the shade of other trees, and so growers have planted native trees alongside hundreds of thousands of coffee trees. There are currently over 500 farmers growing coffee on Mount Gorongosa, many of them women. For these coffee growers, farming has come a long way from unsustainable slash and burn agriculture.

Success stories like this mean that Greg Carr has many enthusiasts – and for good reason. Without him, it is unlikely that Gorongosa's transformation would have been possible. However, a foreigner managing a public resource raises dilemmas. Who should get to decide what the landscape of the park should look like and what people can do in it? Greg employs Mozambican rangers and conservationists, but these are still the privileged few, chosen by the park's managers. Giving a voice to the least powerful members of society and ensuring their needs aren't neglected is a much harder proposition, but involving local people in decision-making is increasingly seen as the appropriate way to run a protected area. People are more likely to comply with restrictions they help create. However, it is extremely hard to do. With 200,000 people in the buffer zone, it is impossible to hear everyone's perspective or to prioritise everyone's needs.

A feeling of disconnection from decisions is one reason some local people are not supportive of Gorongosa and see the park as having been sold to foreigners. But the Mozambique government has little alternative to working with conservation partners – and it also seems fair that ecological restoration *should* be funded by foreigners. We all benefit from services such as climate regulation, so why should we expect the Mozambican government to foot the bill? Our lives are enriched simply from the knowledge that places like Gorongosa exist. That is surely worth paying for.

The affinity that so many people feel with Africa's majestic landscapes could allow the Gorongosa Park to become self-sufficient, using tourism income to end its reliance on international donors. This is Greg's vision, and so infrastructure has been expanded to include facilities such as a luxury tented camp. Tourism gives local people the chance to make a living from the land without degrading it. Even for those not directly employed, Gorongosa is better than most at ensuring that income benefits a range of people. It also follows Mozambique law, which states that every conservation area must give 20 per cent of its revenue to local communities.

Reliance on tourism is not without its risks, though, as the Covid-19 pandemic has shown, particularly in a country with a history of security concerns. Nor will even a self-funded park run by Mozambique nationals end the Western influence on how it is managed: tourists will only come if the park meets their expectations of a 'wilderness' landscape alive with animals and generally empty of people. This has been abundantly clear in the way that the park is managed – one early agreement stipulated that fishing was allowed only on parts of the lake not visible to tourists.

These aren't the only challenges for the future. At the moment, development projects are trying to bring people above the poverty line, but what happens once this is achieved? Will communities accept the inequality between local Mozambicans and the Western visitors to the park, or will they aspire to a more affluent lifestyle? Can that be achieved without increasing their environmental footprint?

If the Gorongosa story has taught us anything, it is that transformations can happen against the odds. If there are honest conversations about the trade-offs and synergies between conservation and human development, these can help navigate the rocky path ahead. A recognition that humans and nature are entwined will help guide national parks towards sustainability and justice.

Navigating the connection between humans and nature at Gorongosa returns us to the Half-Earth debate. Banishing humans from half the planet and then giving people free rein on the other half is clearly not a fair or sustainable solution to the problems we face. And Half-Earth will fail if we can't manage the land we live on sustainably, whatever proportion of the Earth's surface that is. So perhaps we need to reframe the question – this isn't about how much of the planet we protect, but what the rules of protection are. It is about having different restrictions in different places, and thinking about who is affected. Some form of protection will be needed on far more than half the planet, but that doesn't need to be blanket bans on human activity. Can we ban mining companies from an area while recognising the rights of Indigenous peoples? Or allow local fishers into marine protected areas while excluding large international fishing vessels? There will be questions such as these specific to each situation, and tackling them will help move the debate forwards.

We need nature to thrive outside protected areas, and humans to thrive within them. The question of how we can ensure that humans coexist with the rest of nature in protected areas felt especially pertinent when I visited Kaziranga National Park in Assam, northeast India.

Kaziranga is famed for its greater one-horned rhinos, huge beasts whose thick skin has deep folds that give it the impression of armour plating. The densities are amazing to witness – my first views of rhinos came from the side of National Highway 37, the main road which cuts through the park, where rhinos can sometimes be seen crossing. The park's incredible growth in rhino numbers is heralded as a conservation success story – a 2018 census recorded 2,431 rhinos, two-thirds of the world's population, yet in 1905 there were fewer than 100. However, while the photos I brought home displayed relaxed rhinos chewing on lush vegetation, they mask a deep-seated conflict.

Although the park is officially designated for nature, it's clear the separation is artificial. Humans and nature are entwined here – from the mynah birds eating parasites on the domestic cows roaming the park's fringes, to the orphaned baby rhinos that were hand-fed by the Duke and Duchess of Cambridge on their tour of India. The tourist vehicles on the roads which criss-cross the park confirm that this land is for humans, not just wildlife. But which humans? Is it me and my husband, tourists who are here to enjoy the wildlife? Or the safari driver who has a 'Wildlife Is My Lifeline' sticker across his windscreen? Or the people in surrounding areas who benefit from Kaziranga's ability to absorb flood waters?

It's easy to imagine the value to all these people as I bounce through the park in an open-topped jeep, racking up an impressive bird list. Some of the park's benefits are physical – the wetlands recharge groundwater, for example, and are important nurseries for fish, which ultimately provide people with valuable protein. Other benefits are less tangible, such as the sense of identity that rhinos give to local residents and to wider society. People around the world value the fact that rhinos exist, even if very few of us can name the five surviving species. Our attraction to animals such as rhinos encourages tourists to come and experience Kaziranga's arresting scenery of open plains, waterholes and forest. Not only does this enrich visitors' lives, it also provides local people with a source of income that doesn't deplete resources. This benefit doesn't reach everyone, though. All the local people who acted as our guides, chefs and drivers were men. If we saw women working in the tourist camps at all, they were bent double sweeping up leaves. As always, the benefits of nature are distributed unequally.

A look at the park's history reveals that many people have paid the price of protecting this World Heritage Site, and many more continue to do so. Kaziranga was initially a game reserve for colonial hunters, but over the last 100 years it has morphed into a strict nature reserve. Its boundaries have been expanding, as the Half-Earth movement would hope, and rules have become stricter. In the core area of the park, local people have been banned from harvesting timber, collecting firewood and using forest produce that was crucial for their livelihoods. Villages were cleared and communities were evicted from their land. This desire to separate people and wildlife has caused lives to be lost – in 2016, two people were killed when they tried to resist eviction.

Other people have died protecting the park and its wildlife. The lodge where I stayed had a board commemorating rangers who lost their lives to drownings, road accidents and killings by poachers or animals. The board ran out of space in 2011, but the deaths continue. Just a week before my visit, a forest guard on night patrol was killed by a charging rhino. Meanwhile, efforts to halt the poaching of rhino horns are, notoriously, not always successful. Sometimes poachers perform extraordinary acts of cruelty, with de-horned rhinos being left to bleed to death. As a result, park rangers have been instructed to shoot suspected poachers on sight, the kind of power usually reserved for soldiers intervening in civil unrest. Such powers leave people divided – do they allow park rangers to protect themselves and the rhinos from poachers armed with sophisticated modern weaponry, or are they simply legitimising execution without trial? In 2015, park guards shot dead more people than poachers killed rhinos. Other people were shot but not killed, and in 2016 this included a 7-year-old tribal boy whose right leg is now permanently damaged.

The park is clearly of huge importance to both people and animals, and unsustainable use of its resources would lead to disaster. However, closing the park to everyone other than guards and a few tourists isn't the only option. Alternative policies could give local people greater access to the park, and allow many people a stake in deciding which path the park should follow. At the moment fringe dwellers aren't consulted about conservation strategies, which often conflict with their traditional ways of living, and so rules are difficult to enforce.

Outsiders – middle-class officials and visitors – currently have more control over the park's fate than the local inhabitants whose livelihoods depend on it. The people who influence the park's management often hold values that are related to belonging, and who they believe belongs in the park is connected to who they believe belongs in India. Kaziranga is in the area of north-east India which extends beyond Bangladesh, and people's desire to cleanse the park of subsistence farmers can get mixed up with an anti-immigrant sentiment focused on Bangladeshi Muslims. It's a reminder that those of us who don't rely on resources from protected areas need to unpick our reasons for maintaining them.

Eco-colonialism was never far from my mind during my trip to Kaziranga, along with the question of whether I was romanticising an inherited idea of wilderness. What did I gain from being in the park? I realised that the separation I valued was not between people and nature, but between nature and modern society. The tangled trees didn't conform to any of society's rules, and it is hard to worry about your place in an office hierarchy while you're watching a feeding trail of ants. The monkeys sitting grooming each other didn't care about my qualifications, my appearance or any of the ways which humans make their status known, so it seemed foolish that I do. On deeper reflection, this seems a healthy message from time spent in nature – in our increasingly regimented world, I was reminded of the irrelevance of many things we value. It's a reminder, of course, that doesn't necessarily need to come from a wilderness where humans and nature are separate, or from a specific past state of nature.

CHAPTER 8

WINNERS AND LOSERS

Salmon vs seals (and tales of trophy kills, badgers and spotted owls)

'Why should we starve so baboons may eat?' This question baffled the farmers living on the edge of Uganda's Kibale National Park. Baboons ate their crops, yet conservationists wanted to protect them. Such sentiments are repeated in different forms around the world, with people questioning why wildlife is viewed as more important than themselves. This resentment is not surprising given that up to 40 per cent of our food is lost to wildlife. Whether this loss is to baboons, or to slugs or fungi in our gardens, there are a multitude of views about how we should respond. Although on the surface conflicts are between humans and wildlife, they often play out as conflicts between different people. The people who benefit from wildlife protection are seldom the ones to pay the price. In every conservation conflict there will be winners and losers among both people and wildlife.

Deep-seated conflicts can go well beyond any material losses, as revealed by the clash over seals in Scotland's Moray Firth. Multiple rivers flow into the North Sea at the Firth's jagged cliffs, creating employment in the ports and in tourism. As well as the whisky, tourists are attracted by the rugged scenery and abundant wildlife. However, the same wildlife that is attracting them is causing problems for the fishing industry.

The Firth is important to both commercial and recreational salmon fishing, though catches are declining, as in so many places. The reasons are complex but the fishing community has placed some of the blame on seals. Seals cause losses in a variety of ways, such as scaring salmon, therefore making them harder to catch. They also eat salmon from inside fishing nets, and damage the nets, allowing salmon to escape. However, the overall impact of seals on the salmon population is hard to quantify, and there is a tendency for people to interpret the evidence based on their world view. Sustainability scientist Dr James Butler, who mediated in the conflict over the seals, explains that 'Seals can cause problems for fisheries, but hysteria has come into this, too – a fisherman will see a seal with a salmon in its mouth and then make assumptions about the number of fish that they eat.'

For decades fishermen had responded to the perceived threat by killing seals, with licences issued by the Scottish government. But in the early 2000s fisheries managers increased their shooting effort, which caused a backlash from wildlife organisations and ecotour operators. It got to the point where fishermen faced the threat that all culling would cease if a solution to the conflict couldn't be found. With no resolution in sight, James led a process of conflict management between the fisheries, conservationists and tour operators. He worked with Dr Juliette Young, who describes the starting point for the debate: 'We had this strange situation where both the salmon and the seals were protected under EU directive, but one was preying on the other. There were very strong and polarised views from the fishermen, the fisheries boards, seal conservationists and the tourism industry.'

Another complexity was that the seals being shot were two different species: grey seals, a common species which is increasing in Scotland, and harbour seals, which were declining. This prompted different opinions even between groups promoting seal conservation. Organisations such as the International Fund for Animal Welfare (IFAW) were concerned about the lives and welfare of individual seals, regardless of the species. Nature Scotland, on the other hand, was concerned about the population status of salmon as well as both harbour and grey seals. To them, the fact that grey seals were increasing was relevant to their decision to license shooting, whereas other people were inherently opposed to killing any seals.

While Nature Scotland viewed both salmon and seals as important, the seals had more supporters – most people don't see a salmon's life as equivalent to a seal's. James describes the deep-rooted views: 'There is a strong cultural memory of images of white seal cubs being clubbed to death in Canada. The image is so emotive that it became taboo to challenge the view that killing seals is wrong.'

With an understanding of such emotions, James began the process of conflict management in an attempt to maintain livelihoods while ensuring a healthy conservation status of both salmon and seals. He brought the different groups into the same room – fishermen who were willing to shoot seals to protect their livelihoods, and conservationists who had a mandate to protect seals. The result of extensive discussions was a management plan focusing on 'rogue' seals. Research showed that these animals habitually came into the rivers and caused most damage, so the previous ad hoc approach to culling was replaced with a more selective response. Juliette explains: 'The management plan was really quite clever because it put stakeholders such as the fishermen, who had felt sidelined, in a prominent position, and it allowed them to have some say over what they were doing. They still had the option of culling seals, but this was much more targeted. From a conservation perspective it was also a win because there were fewer seals being shot.'

This assessment of the plan's success is strikingly different to that of traditional conservation. Rather than crediting the science about wandering seals as the reason a solution was found, Juliette focused on the human interactions. This change in perspective happened early in her career, when she realised that conservation conflicts wouldn't be resolved simply by collecting enough ecological data. She is now a biodiversity policy researcher at the Centre for Ecology and Hydrology in Edinburgh, but her early work included chimpanzee rehabilitation in Sierra Leone. She repeatedly saw problems arising when people had different views of wildlife and how it should be managed, and this led her towards a PhD in social science. The transition required her to look at things in a whole new way.

She now sees that the perception of power is an important aspect of conservation conflict: 'People often have strong views about what they feel the power dynamics are. Some conservationists have a perception that fishermen have quite a lot of power when it comes to decision-making, whereas fishermen themselves might think that they don't. One group can feel that they have no say, no voice, yet they aren't viewed in that way by others.'

She has found that sharing these perceptions can be really useful, helping to raise empathy between different groups. It is also important to listen to people's broader views, as we have a tendency to assume we know what other people believe. 'People can be so entrenched in their world view about the other stakeholders that it's impossible to move forwards,' Juliette says. 'They often have a very clear idea of what they think others' views are, even if they've never even been in contact with that group of people.'

James describes an example he had encountered – the assumption that anyone who killed a seal wanted to do so. In his experience, people seldom want to kill an animal. The motivation for killing seals in the Moray Firth was to protect fish stocks, not because killing was enjoyable or cathartic.

Our different views can be very basic, to the point that we are not even aware of holding them. Some of us see the natural world as fragile and consider ourselves as stewards of nature. We may see it as logical to conserve nature by setting up reserves which are out of bounds to everyone other than people who share our deep appreciation of nature. Or we may view nature as much more resilient, which will affect the way we believe nature should be managed. It is often a product of our own situation. As James explains: 'The ability to think about conservation and to question issues such as seal culls is a luxury. The closer you live to the ground, the greater the necessity to make a livelihood from natural resources.' The irony is that those of us who have the economic security to devote attention to seals are often leading more environmentally damaging lives. Many people who disapprove of seal culling also eat salmon, and their choices about diet and travel, for example, mean they may have a greater impact on wildlife than the small-scale fishermen do.

Throughout this book, I have argued that there are no right answers – nobody has a trump card that dictates the 'correct' solution. The discussions that James and Juliette facilitated are a reminder of how important this is. If we all come to the table believing we have the answers, we risk a perpetual stalemate. That approach is likely to make sure the winners are the most powerful people, and probably the most charismatic wildlife. If we enter debates with an open mind, along with an appreciation of the value of all humans and sentient animals, we are likely to come out with fairer, more sustainable solutions.

The dynamics of power are particularly challenging in the impassioned debates about trophy hunting. Wild animals are hunted in a huge range of situations, yet the harshest judgement falls on only a tiny minority of the hunters, and most of all on Minnesotan dentist Walter Palmer. On 1 July 2015, Palmer fired his crossbow and wounded Cecil the lion with an arrow. Along with a local guide, he then tracked Cecil for eleven hours before shooting him again. This second arrow marked the end for 13-year-old Cecil, a black-maned lion from Zimbabwe's Hwange National Park, but was the beginning of an explosive international debate. Cecil was a major tourist attraction and a lion with a name, and his death sparked deep emotional reactions. Social media exploded in debate, and campaigns sprang up. The public outcry led to change, with some Western governments making it harder, or even impossible, to import lion trophies.

The intense backlash was partly because of Cecil's beauty, but also because of the motivation behind his death. We accept battery farming and abattoirs in the UK, yet the desire to kill for glory is so hard to imagine that for many it feels depraved. Photos of trophy hunts are certainly hard to witness. Cecil wasn't the first lion that Walter Palmer had shot, and pictures circulated online of him grinning as he posed behind a previous trophy. These images help us to separate our own lifestyles from Palmer's – the press sees no contradiction when it informs its meat-eating readers that 'we must stop cruel and immoral trophy hunting'.

There are of course some genuine differences between trophy hunting and meat eating, and these are reflected in the views of some of its high-profile critics. British explorer Sir Ranulph Fiennes, for example, states that stopping these hunts would recognise the destruction that Europeans inflicted on wildlife in their former colonies. To Fiennes, trophy hunters are 'bullying bastards'.

Others see the power of Western lobby groups as decidedly sinister – their influence may be continuing colonialism, not atoning for it. A British review of trophy imports came in the wake of Botswana's 2019 decision to end its moratorium on trophy hunting, and it raised the question of whether the British government should have any influence over trophy hunting in other countries. Hunting – of deer, rabbits and ducks – is common on British soil. We even release millions of non-native pheasants and partridges into the countryside each year specifically so they can be shot. It therefore may seem a strange contradiction that we dedicate so much energy towards trying to reduce hunts overseas, while sanctioning them on our soil.

As things transpired, the UK announced only a partial ban on trophy imports in 2021, much to the outrage of campaigners. In reality, it's not clear whether even a complete ban would reduce trophy hunting now that hunts are photographed and often filmed. In fact, the opposite may occur. Taxidermy is expensive, so by removing the pressure to bring back trophies the ban could make hunts cheaper and therefore attract more people. However, this point is rarely considered, and people on both sides of the debate see trophy import bans as a way for Western countries to exert control over what hunts take place overseas.

Philda Kereng, Botswana's Minister of Environment, Natural Resources Conservation and Tourism, therefore asked the British government to 'take into account the interests of those most affected by the proposed ban of hunting trophies'. She highlighted that, while trophy hunting can provide cash, employment and meat, in its absence elephants and predators pose a greater threat to rural livelihoods. Botswanan civil society organisations and citizens, who are overwhelmingly in favour of the government's decision to allow trophy hunting, are unable to keep up with the international social media campaign against it.

So who is right? It can be comforting for people in the West to believe they can contribute to the protection of endangered species with just a few clicks, making a donation or signing a petition, but the reality is extremely complex. Some of the strongest supporters of trophy hunting are conservationists, who argue that the money which flows from it supports conservation and local livelihoods. Palmer reportedly paid $54,000 to hunt Cecil, and this kind of money can make it worthwhile to protect wildlife habitats. Although trophy hunting articles in the press often cite outrage from conservationists, these tend to be from groups such as Born Free that focus on animal welfare. Conservation organisations such as the Worldwide Fund for Nature (WWF) and the

International Union for the Conservation of Nature (IUCN) believe that trophy hunting should be considered when it can bring both conservation and community benefits.

One reason for the polarised divide is that no two situations are the same, so it is easy for an interest group to cherry-pick cases. For a start, not all trophy hunts are legal, so some don't follow regulations that protect animals and people. Most concerning are 'canned' hunts of animals specially bred for the purpose. As you can imagine, standards of welfare can be very low for species such as lions reared in captivity, with crowded conditions and insufficient food and water. In the worst cases they are illegally drugged for the hunt.

Animals hunted from the wild generally have a much higher quality of life, but may not experience a painless death, as the case of Cecil the lion shows. And even painless killing can cause suffering for animals that remain – for species such as elephants that form close social bonds, the loss of one animal can have a profound effect. However, this can be weighed against the alternatives. If money from trophy hunting protects habitat, many animals will benefit. Money could come from other sources, such as photo tourism, but people pay a lot less for this and it too has welfare concerns. Elephants used for safaris are often kept in very poor conditions and taking them from the wild has the same impact as trophy hunting on the elephants left behind. Even for jeep safaris, tips are often highest if visitors get the best photos, making a powerful incentive for drivers to get too close to animals, and sometimes feed them.

Jeeps are often carefully positioned to ensure a background of open plains, keeping the cluster of vehicles and giant lenses out of view. Recent work shows just how damaging this can be. In Kenya's Maasai Mara, for example, cheetahs raised an average of just 0.2 cubs per litter in areas with high vehicle densities between 2013 and 2017, less than 10 per cent of the number raised in areas with low tourism.

Not only can trophy hunting attract funds for conservation; it can also protect local people. Although the number of elephants shot in Botswana was very low, the fear of hunting had kept them from roaming far from protected areas. The moratorium on hunting therefore put them in conflict with villagers who reported that elephants destroyed their crops, scared them and damaged trees. Increased human–wildlife conflict was one of the reasons Botswana gave for lifting its hunting moratorium in May 2019. This was a reversal of previous laws that had been both applauded by Western animal protection organisations and condemned for their impact on local people. The laws had been criticised for removing rural people's rights to use their natural resources, and introducing a militarised approach to wildlife management which alienated wildlife from rural citizens. After lengthy consultations, Botswana has restored ownership to local communities, giving them a stake in the management of wildlife on their land and enabling benefits to be shared.

Of course, trophy hunting is by no means the only way to manage animals that damage crops and attack livestock. Deterrents such as beehive fences can protect crops from elephants, or contraception can keep down the numbers (as elephant herds are much smaller than herds of deer, for example, this option could be feasible). These interventions have costs that may be beyond the means of subsistence farmers, in which case they will only happen with outside help. The drama of a campaign against trophy hunting may attract more attention than a project that addresses human–elephant conflict, but focusing on alternative approaches may be a more powerful way to reduce human and animal suffering than shouting about trophy hunts.

There are also ways to reduce the negative impacts of trophy hunting. From both a conservation and welfare perspective, it makes a difference which animal is shot. Not only is the species relevant, but also the condition and status of the individual animal. Large male lions may be particularly attractive to trophy hunters, but can be the worst choice from the perspective of the pride. Their loss can affect social groups and reproductive success. If weaker animals are taken instead, they may have a less painful death than they otherwise would have done. There are also regulations aimed at preventing kills of mothers with dependent young, including closed seasons for hunting.

Ultimately, trophy hunting has potential to benefit people and wildlife, but this won't automatically happen. It will bring benefits only if it is well regulated to ensure populations aren't put at risk, animal welfare guidelines are followed and income goes to conservation and local people. Major barriers to this include corruption, and funds can end up lining the pockets of officials and tour operators. Given the complexity of different situations, it may be more effective to fight for legislation that protects people, habitats and animal welfare rather than pressure for a blanket ban on trophy hunting. This won't be easy, and will require consultations with local people. However, fair and effective legislation seems like a worthwhile result.

Cultural divides also occur over which species should be protected and which should be persecuted. Our innate reactions to certain species can cause us to overlook the important roles they play, as we saw earlier, and can bring us into conflict with people who view them differently. Scavengers, for example, often elicit feelings of disgust. It's no accident that the hyenas in *The Lion King* are in league with the enemy, and blowflies are never welcome guests. However, such reactions are based on misunderstandings. Spotted hyenas are in fact excellent hunters, and lions are just as likely to be scavenging their kills as the other way round. When they do scavenge carcasses, this helps prevent the spread of disease, and scavengers such as flies likewise play a vital role in recycling nutrients.

Despite their importance, many scavengers face persecution not protection, with vultures, for example, being regularly shot or poisoned. Their bald heads and plain, shaggy feathers mean there's no denying that they don't meet human ideals of beauty – Darwin described vultures as disgusting. Not everybody agrees with this assessment, which is fortunate given that over half the world's vulture species are threatened with extinction. Throughout history, vultures have left people divided. In ancient Egypt, vultures were a symbol of maternity, partly because they were believed to all be female.

Still today, vultures divide people. They are often recognised for their effective sanitation service, and in Tibet are considered to be sacred. They play a central role in sky burials, in which human bodies are left out for vultures and other scavengers. These funerals used to take place in Mongolia, Bhutan, Tibet, China and India but are becoming increasingly rare, partly due to declining vulture populations. One cause is the toxic impact of the anti-inflammatory drug diclofenac being used on cattle. The drug has inadvertently poisoned millions of vultures eating livestock carcasses in Asia and beyond, causing some populations to plummet by over 99 per cent since the 1990s. Deliberate persecution also plays an important role in the declines, partly due to the mistaken belief that vultures predate livestock. It's a pervasive problem – when a bearded vulture turned up in England's top grouse shooting area in 2020, the Royal Society for the Protection of Birds (RSPB) rated its chances of being killed as so high that they warned its presence there was like 'a turkey spending Christmas at a butcher's shop'.

These different responses to vultures reveal how our views are shaped and reinforced by culture, and how they can vary even within communities. Species that are widely viewed as important in the West don't necessarily have the same support elsewhere, which serves as an excellent reminder to analyse our intuitive preferences. The cultural attitudes we've all absorbed aren't universal, and some of our basic assumptions about what's important may not be shared. This is perfectly illustrated by the different reactions to orangutans. All three species of orangutan are endangered, and they are a focus for conservation charities such as the WWF and the Wildlife Conservation Society. However, the habitat destruction and deforestation that cause orangutan declines have also pushed them closer to human activity, leading to conflict.

Orangutans eat crops and cause people to feel threatened, so they are often chased away, wounded or killed, and sometimes even captured and beaten. In contrast, local people prize certain species that can be overlooked by foreigners. Villagers in Borneo, for example, often see fish as being far more important than orangutans, particularly because they're a source of protein and income. Green leafbirds are also seen as more important than orangutans, and hornbills as roughly equivalent. This is partly because green leafbirds are sold as songbirds in local markets, and hornbills are important in Dayak culture. In other ways, local priorities may match the ambitions of foreign conservationists, though not always for the same reasons. For example, protecting forest habitat for orangutans can lead to cleaner air and water, and this is where local people witness the benefits. We all have different outlooks, and define success in a different way – understanding this will pave the way for frank discussions, bringing diverse voices into the conversation.

Just as conservationists and local people may see things differently overseas, the same can be true at home. Conservationists often focus on protecting rare birds, for example, yet plenty of people haven't even heard of these species. What many of us value most are the common species we hear singing. The chances are that you have taken far more pleasure from the blue tits and house sparrows in your garden or local park than the tiny red-necked phalarope population breeding on Fetlar in the Shetland Islands. These elegant birds are unusual in that it is the males who raise the young, and because they are long-legged wading birds with webbed feet, so they can swim like ducks. They are a wonderful sight and I feel privileged to have seen them, but even so my life has been more enriched by the common birds around me. The common species I enjoy often thrive in urban environments that are less diverse than the 'pristine' habitats that conservationists value so highly.

The urban habitats that city dwellers appreciate may be considered degraded by conservationists, and trees valued by local residents may be looked down upon for being non-native. However, as we have seen, we need to question the assumption that nature is automatically superior if it resembles past ecosystems and contains species which arrived in the region long ago. Likewise, the brownfield sites that most of us dismiss as wasteland are often home to a great diversity of life. When we import topsoil and plant grasses to convert these sites into green spaces, we eliminate some of the organisms that exist there. The attractive wildlife of parks can be less diverse than the ever-changing life in brownfield sites. This doesn't mean the conversion of brownfield sites is wrong, but a habitat that is home to a greater variety of species or genes isn't necessarily superior.

The acknowledgement that conservationists cannot be guided by an objective goal is important. We conserve what is valuable to us, and often exclude or neglect people of different backgrounds, meaning that the nature they value won't be protected. This is disturbingly clear in cities – studies in America show that neighbourhoods with a high proportion of African American and Hispanic residents have fewer trees. Likewise, there are fewer parks and public green spaces in poorer and non-white neighbourhoods. This is particularly worrying given the benefits of nature in cities, from reducing high temperatures to improving mental health. Even viewing nature from a window can reduce stress. If we took environmental justice seriously when allocating conservation resources, we would create greener cities that benefit the physical and mental health of people who lack the access to nature that so many of us enjoy.

In order to realise the full benefits of conservation, we need to involve diverse people in decisions about the wildlife around them. There are often barriers to hearing different voices, but these can be overcome. London's Natural History Museum, for example, is focusing on policies to increase diversity among its staff and to review the way it presents information about its specimens, acknowledging the collection's colonial history. It is also running an Urban Nature Project to improve its wildlife garden and engage a broader audience in its design. From consultations with the neurodiverse community to LGBT+ tours, the museum is proving that far more can be done to welcome diverse perspectives. Everyone has an impact on wildlife, and everyone's well being is affected by it, yet decisions about the natural world have so often been made by the privileged few.

Britain's badger cull provides a good example of where a cultural divide even between people living in the same village can seem insurmountable. It's a long-running debate. Badgers have been implicated in the spread of bovine tuberculosis for almost five decades and we're still locked in the strange contradiction of a government spending millions of pounds culling a species they have granted legal protection to.

Bovine tuberculosis (bTB) is caused by *Mycobacterium bovis*, a close relative of the bacterium which causes human TB, *Mycobacterium tuberculosis*. It's unusual in infecting a range of hosts, including badgers, deer, cats, cattle and, crucially, humans. The bacteria can pass to humans in infected meat or milk, although meat inspection, milk pasteurisation and cattle testing have dramatically reduced this. As a result, fewer than fifty human cases are diagnosed a year in the UK, and these rare cases are treated by our healthcare system. The impact on human well being, however, reaches much wider than the disease itself. Bovine TB in cattle can place an extreme burden on farmers, if their cows are among the 30,000 slaughtered each year in the UK following positive tests. Not only is this an emotional strain, but it can also cause financial difficulties – government compensation is available, at great cost to the taxpayer, but it may not be enough in an industry with such low profit margins.

This creates a huge imperative to reduce the transmission of bTB, which doesn't just occur between cows. The link between badgers and bTB in the UK was made in 1971, when a dead badger was found on a Gloucestershire farm experiencing an outbreak. The government acted fast, and introduced large-scale culls, killing badgers by pumping hydrogen cyanide into their setts. This was immediately followed by controversy, and a review concluded that 'gassing' doesn't kill the badgers fast enough to be considered humane. A range of other techniques were tried in the 1980s and early 1990s, but still the evidence about the effect of badger culling on bTB was circumstantial. This could only be rectified with a large-scale experiment, and so the Randomised Badger Culling Trial (RBCT) began.

After almost a decade of research at a cost of £50 million, the trial concluded that 'badger culling could make no meaningful contribution to bovine tuberculosis control in cattle in Britain'. You might expect this to end the debate, but not everyone agreed. Some scientists and vets challenged the conclusion, and farmers' attitudes remained remarkably unchanged. One reason for the dispute was that, rather than showing no effect of badger culls, the trial found that culls sometimes reduced bTB in cattle and sometimes caused it to increase. The decreases tended to occur in the core areas of the cull, whereas increases occurred around the edge. This increase is believed to occur because reducing badger numbers disrupts social groups and causes survivors to roam more widely, spreading bTB.

The trial may not have solved debates about the cull's effectiveness, but one thing it made clear was that the method used was too expensive. Badgers had been trapped in a cage before shooting them or delivering a lethal injection, which is labour-intensive. When the government returned to badger culling in 2010, they instead began trials of free shooting badgers without catching them first. This raised serious concerns about welfare, and pilot culls failed to meet the government's target of fewer than 5 per cent of badgers taking more than five minutes to die. This provided further motivation for protesters, who took to the streets and to social media, and physically disrupted culls. Widespread criticism also came from scientists who doubted the cull's humaneness or effectiveness.

One reason for the continued culling, which still goes on today, is the lack of an alternative solution. There is a BCG vaccine available to protect cattle, but it isn't completely effective, and surveillance becomes very difficult because vaccinated cows often test positive for bTB even when they're not infected. Developing a test to distinguish between infected and vaccinated cows is therefore a priority if cattle vaccination is going to prove a feasible answer. Badger vaccination also shows promise, though current vaccines don't give complete protection against bTB. As well as preventing the need for shooting badgers, vaccination has the advantage of being cheaper than culling, thanks to the support of volunteers – the Wildlife Trusts, for example, assist with vaccination in the hope of preventing culls. The UK government has therefore announced that intensive badger culls will be phased out over the next few years, in favour of vaccination and surveillance. This was welcomed by conservation groups, but we are yet to see whether it will happen. In 2021, 75,000 badgers are set to be culled, and the cull extended to several new areas.

Meanwhile, the debate continues about whether badger culls reduce bTB. The evidence is sufficiently complex for both sides to believe their arguments are supported by science. The government website quotes a 2019 study as showing that intensive culls have been associated with reductions in bTB. The study had in fact investigated three areas where culling was taking place, one of which saw an increase in bTB. This third area got no mention on the website, nor did the fact that bTB was already decreasing in the other two areas before the culling started.

Not only are there reasons to mistrust the way that data is interpreted, we could also question why the bTB debate revolves around culling badgers. Alternative ways of controlling the disease get much less attention, even though bTB is spread far more by cattle than by badgers. It seems that 90 per cent of transmission occurs directly between cattle. A 2018 review commissioned by the government concluded that the culling debate detracted attention from what can be done on each farm, and sidelined issues such as the transportation of high-risk cattle.

On the other side of the debate, it's worth considering why badgers get so much support. In many ways, they are an unusual focus for conservation. The European badger is a common, widespread species, occurring right the way from Spain to Iran and up to the Arctic circle. It's not declining; in fact, quite the contrary. The little evidence we have about numbers in the UK suggests that the population has grown in recent decades, reaching an estimated 485,000 in England and Wales. This is a far higher density than in other parts of Europe, although populations seem to be largely stable or increasing elsewhere too. People are, quite rightly, concerned about the welfare of wild animals, yet, as we saw earlier, we kill less attractive species with little thought. Badgers, it seems, are viewed differently to rats and mice.

The fixation with badger culling, and the fact that the evidence is interpreted differently by farmers and conservationists, suggest something deeper is at play. Core to this debate is our relationship with the natural world, and how that shapes our personal identities. The badger's striped face has long been an iconic part of the British countryside, appearing on coats of arms and in place names. The place name 'Broxbourne', for example, is believed to come from the older word for badger, *brock*, and the Borough of

Broxbourne logo still depicts a badger. It is therefore hardly surprising that emotions run high when considering the large-scale killing of an animal that has shaped our identities.

However, badgers haven't always enjoyed such support. In Tudor England, they were classed as vermin and fetched a generous bounty of 12 pence per head. Their meat was eaten and badger hunting was a popular sport, with dogs such as dachshunds bred to enter setts and fight with the occupants. Although the bloodsport of badger baiting was banned in 1835, it remained common (sadly, even continuing today), and only towards the end of the century did the image of badgers transition from celebration of hunting to concern for animal welfare. Badgers came to be portrayed as clean, civilised, sociable and family-orientated, partly because they rear families in the same area for generations.

Predictably, not everybody's attitudes were quick to change – complaints were still raised about the damage badgers do to crops, and they were accused of taking chickens. The elite were also concerned about their impact on foxes, which ironically they wanted to protect for hunting. The badger's habit of digging young rabbits from their burrows and eating them was still used by some people to counteract the 'good badger' image, whereas others lauded this as a pest control service. Overall, their positive image was winning out, though, and the badger's transition from vermin to valour was sealed by Kenneth Grahame's 1908 novel *The Wind in the Willows,* in which Mr Badger is intelligent and wise and a fierce defender of his friends.

The different portrayals of badgers are visible in news articles today, with the 'bad badger' making a comeback with bTB. Supporters of the cull often portray badgers as disease-ridden and destructive. In such instances, badgers tend to be discussed in the plural, making it easier to highlight their high numbers rather than their welfare. Elsewhere, the sociable, good badger is portrayed as an innocent victim. Our attitudes also reflect a relationship with the land. Encountering a wild badger can bring a sense of connection to nature, and this love of the British countryside is threatened by the knowledge that we kill them. However, not everyone's feeling of connection to the land comes from the same source. People opposing the cull may see themselves as protectors of wildlife, and portray farmers as motivated by profit. Farmers, on the other hand, may consider themselves the custodians of the countryside.

None of us are immune to the tendency to fit evidence around our values – our brains have cognitive biases that cause us to resist evidence that challenges our world view. And we all have a need to be heard and understood, so anyone who considers themselves to have lost out due to conservation is more likely to respond to empathy for their experiences than to data claiming they are wrong. A farmer's fear of bTB and their feeling of loss if cows test positive are real, even if badgers are only the scapegoat.

In conservation terms, the badger cull debate has been surprisingly narrow, with its wider ecological impact rarely considered. We know remarkably little about what the effect could be, to the point that the years of culling haven't been accompanied by an Environmental Impact Assessment (an oversight the government has been taken to court over). We do know, however, that hedgehog numbers increased during culling trials, something conservationists could have chosen to celebrate. Badgers prey on hedgehogs, and may compete with them for food. They also compete with foxes for food, and culls seem to increase the number of foxes.

Information about the cull's broader impact will no doubt sit differently with different people, simply because we all relate to the countryside in different ways. That's no bad thing; in fact, it is something to celebrate. However, it's a reminder of how our outlook can alter our judgement. In reality, branding badgers as over-abundant pests or as a quintessential part of the countryside has no effect on their capacity to suffer, their impact on wildlife or their role in spreading bTB.

In conflicts between people and charismatic animals, the framing of Britain's badger cull as rural versus city folk is a familiar story, as is a failure to consider the full context. Sidelining the wider ecological and human consequences created a bitter divide in the debate over North America's northern spotted owl, which has been rumbling along for decades. The early conflict was catchily summarised by *Time* magazine's cover story on 25 June 1990: 'Owl vs Man: Who gives a hoot?'

Spotted owls are distributed in a band down the West Coast of America, from Canada to Mexico, as well as pockets reaching into Utah and Colorado. The species as a whole is declining and is listed as 'near threatened', meaning it is on its way to becoming endangered. It is divided into three subspecies, and the northern subspecies is found along the coast from north California to British Columbia. These elegant hunters are an average of 43 centimetres tall, almost up to my knee, and their dark brown wings are decorated with white spots. Their population fell in the twentieth century, and in 1990 the northern spotted owl was listed as threatened under the US Endangered Species Act.

Two of the primary reasons for the owl's ongoing decline were timber harvest and forest fires. The first recovery plan therefore focused on forest protection, with new regulations preventing timber harvest in critical habitat. The drastic reduction in timber harvests on federal lands hit hard in the Pacific Northwest. Employment in the timber industry plummeted in the 1990s and the owl was viewed as the culprit. Although there's debate about the extent to which the owl was to blame, an analysis by the US Forest Service concluded that spotted owl protection caused the loss of 11,400 timber industry jobs in the decade. The controversy pitted individual loggers and small sawmill owners against environmentalists. Creative bumper stickers appeared to support the loggers, with messages such as 'Kill a Spotted Owl – Save a Logger' and 'I Like Spotted Owls – Fried'. In Oregon sawmills, plastic spotted owls were hung as effigies.

The environmentalists' victory turned out to be hollow, however, as the owl population kept declining. Barred owls were rapidly expanding their territory westwards, and scientists realised that competition between the two species was far more of a threat than anyone had understood. Barred owls have higher annual survival and rear more chicks, so their populations can quickly outnumber spotted owls. They have now extended their range to completely overlap with the northern spotted owl, and occur at higher densities throughout the whole area. The two species consume similar prey and barred owls defend their territories aggressively. They have even been known to kill spotted owls. The realisation that the threat was another owl caused some conservationsists to became entangled in a new controversy: should we remove the barred owls?

Evidence suggests that culling barred owls is a potential solution, with trials in Canada looking positive. Canada's population of 1,000 spotted owls has declined to around thirty, and half of those are housed in expensive captive breeding facilities. Preliminary results from the removal of more than 150 barred owls indicated that this was beneficial for spotted owls. Trials of barred owl removal in the USA have also been promising, allowing spotted owl populations to stabilise. This leaves many of the owl's supporters with a dilemma – it's one thing to argue against destruction of the forest, but it's a very different prospect to argue against owls.

Halting the decline of the northern spotted owl and protecting this subspecies from extinction will come at a high price for both barred owls and taxpayers. The trials alone have cost millions of dollars and, even if barred owls were completely eliminated from certain areas, it's very likely that they would quickly recolonise. If we are going to protect the spotted owl from the barred owl, culling would need to be maintained year after year. We don't yet know how extensive barred owl removal would have to be in order to be effective, but studies suggest that culling rates would need to be very high – many thousands of barred owls would need to be shot. Is this a price worth paying?

Perhaps it doesn't matter that the forests of western USA are home to pale owls with dark bars down their breast, rather than darker birds with white flecks, even if this was facilitated by human-made habitat changes. However, we still need to consider practical impacts. The barred owl is ecologically similar to the northern spotted owl, but not completely equivalent. Barred owls have more diverse diets, so by replacing the spotted owl they could cause wider changes to the food web. They also occur at higher densities than spotted owls did even at their peak, so consume more prey. Some species are therefore likely to lose out.

There's no intrinsic reason to ensure the northern spotted owl doesn't go extinct, as we saw earlier. However, losing genetic diversity in this way can have unknown impacts. For example, we don't know how different species and subspecies would fare if a new disease began to spread. Who knows what the future may bring, and which species or subspecies would thrive under new conditions, or what impact they may have. This argument is more powerful when we consider species, as they are more likely to play different roles, but it still applies to subspecies or genetically distinct populations.

For some people, the spotted owl may simply be a hook to attract support for forest conservation. It's certainly a strong motivator – we all benefit from services (notably carbon storage) provided by old-growth forests. However, it's a reminder of the need to be clear about the ultimate objectives. If this is really a man vs forest debate rather than man vs owl, then maybe the arguments should be different. Every story has winners and losers among both wildlife and people.

EPILOGUE

SHAPING OUR FUTURE

Reasons to be cheerful

When Covid-19 spread around the globe in 2020, the way we had been living our lives suddenly became impossible. But, alongside the tragic consequences, the pandemic has revealed our incredible adaptability and our willingness to make sacrifices. We gave up freedoms in order to save lives, from international travel to visiting our families. For many people, though, this upheaval wasn't an isolated event. Lives have already been transformed by changes to the natural world, whether this is from wildfires, collapsing fish stocks or rising sea levels. The burden has so far been predominantly borne by poorer nations, but wealth won't protect us forever – we can't expect to keep depleting the biosphere without feeling the effects. We therefore face a choice. We can wait until change is thrust upon us, as with the pandemic, or we can start the transformation now. We can take inspiration from our ability to change so completely, and think about the future we want to build.

As we've seen throughout this book, protecting nature can play a huge role in creating this better future. Nature sustains our lives in so many ways, from providing food to calming our stress, and our current trajectory is putting all these benefits at risk. In order to create a future where humans and the rest of nature thrive, we are likely to need transformational changes in society. These will come not just from individuals but also from businesses and governments, and will take many forms. We need everything from small-scale conservation projects to global agreements.

The situation is serious. Nature faces an onslaught from many directions, and some declines are irreversible. The International Union for the Conservation of Nature (IUCN) has identified over 24,000 species that face a significant threat of extinction. The prospect of reversing this trend can seem overwhelming, and it's easy to get trapped by a feeling of helplessness in the face of all that has to change. However, pessimism risks being a self-fulfilling prophecy, and can blind us to the many positive signs. If we instead focus on the reasons for optimism, we can see sprigs of hope all around us.

Perhaps most encouraging is that more and more young people from around the world are taking up environmental causes. High-profile cases show what can be done, with prominent examples including Greta Thunberg's Fridays for Future movement. Look closer, and there are countless examples of people working at smaller scales, changing attitudes and behaviour in their own community. Teenagers in the Philippines are encouraging people in their neighbourhood not to hunt wild birds. Villagers in Indonesia are seeking support to release the monkeys caught in traps they had set for wild pigs. Children in my village are looking for ways to reduce their school's carbon footprint and waste.

There are also signs that conservation is having an impact. Forest decline is slowing globally (albeit not fast enough, especially in the Amazon) and temperate forests are actually increasing in area. Many species that had previously been persecuted are making a comeback, ranging from sea otters to bald eagles. And stories of hope aren't restricted to such charismatic species. Despite conservation's continued bias towards popular species and habitats, the unloved are increasingly getting a look in. Recent work on parasitic plants reveals that even they have some supporters.

About one in ten flowering plants are parasites. Some, such as mistletoe, retain the ability to photosynthesise, whereas others derive all their energy from the host plant and have lost their green pigment entirely. A beautiful example is New Zealand's woodrose, also known by its Maori name of *Pua o te Reinga*, 'flower of the underworld', which alludes to the way its flowers emerge from below ground. The parasite attaches to the roots of a tree and all that appears above ground is a dense white flower protected by thin brown leaves. It isn't just the flower that is known for its beauty. The name 'woodrose' comes not from the plant itself but from the structure it induces in the host tree: the root grows into a fluted disc which looks a bit like a carved flower. Sadly, this hidden structure has contributed to the parasite's serious decline, because the plants were dug up so the host's attractive root could be sold.

It's not all bad news for the woodrose – the plants are doing well in some areas. However, their recovery is limited by the fact that seeds are seldom spread more than a few metres. We're not quite sure why, and it could be a shortage of animals who eat the tiny fruits and spread the seeds. The best option is therefore for them to be spread by human animals, so scientists in New Zealand ran trials to see if they could expand the range of the rare woodrose by spreading seeds to new areas.

This task is far more complex than scattering wild flower seeds, and scientists initially assumed it would be difficult for a variety of reasons. In particular, the parasitic plant needs to find the root of a host tree. Translocations of parasites always come with the challenge of uniting them with a host, and this is made more of a problem if we don't know what their preferred hosts are. In the case of woodrose, the host can be hard to identify because host and parasite meet underground where their connection is hidden from view. It's a question the study shed some light on − although the woodrose can parasitise a range of trees, it established particularly well on the evergreen shrub kōhūhū and on lancewood trees. Another specific requirement is that it has a close relationship with its primarily pollinator: short-tailed bats. The woodrose flowers for just two to three weeks in early autumn, and its musky scent attracts bats to drink its nectar and spread its pollen.

Despite these challenges, trials were successful and the woodrose became established in new areas. It's another sign of hope, not just for the woodrose, but also that translocation will work for related species too. This is how momentum for conservation builds − the more we learn about one species, the more knowledge we have to apply to others. It also touches on the issues discussed earlier about the freedom to make choices on where a species should be without ideological constraints based on where it has been found in the past. The current focus is on transporting woodrose seeds to areas where they previously appeared, but the future possibilities go far wider. Having freed ourselves from the shackle of believing that species 'belong' only in their past ranges, we open up possibilities for assisted colonisations. In a changing climate, translocations to new ranges may be the only option for some species. Our expanding knowledge makes this increasingly feasible, while we have learnt from past mistakes about the need to carefully consider any possible effects. The consequences are now considered very carefully, making translocations an exciting prospect.

Perhaps the greatest reason that woodrose conservation gives us cause for optimism is that the ongoing project is a collaboration between Maori and other conservationists. This reflects a wider shift towards a more inclusive form of conservation. The journey is by no means complete, but there are many examples of conservation transitioning from a colonial past to a future which embraces the needs and views of local people. This is true even in the developed world, where movements connecting people of all backgrounds with nature include groups such as Flock Together, a birdwatching community for people of colour. The collective began with outings in London and is now spreading to cities from Tokyo to Toronto. As well as tackling the structural racism that has kept people of colour out of green spaces, such organisations can facilitate environmentalism. Some members collect ornithological data, and hope to protect green spaces by blocking construction plans. By bringing more people into conversations about nature, we can help build a new future that everyone wants to see.

More reasons for optimism come from the opposite end of the spectrum, emerging from global agreements rather than local actions. The IUCN is constantly reassessing the conservation status of the species on its Red List of Endangered Species, and recategorising them based on their extinction risk. All too often these re-categorisations show a negative trend, but not always. The Guam rail, as we saw, made the ultimate move from 'extinct in the wild' to 'endangered', and other species, too, have moved into more favourable categories. Mountain gorillas, echo parakeets and the Australian trout cod are just a few of these success stories that show the power of conservation. In the case of the fin whale, its reclassification from endangered to vulnerable reflects the power of a global effort to protect whales from extinction.

Growing up to 27 metres long, fin whales are second in length only to the blue whale, and are easily distinguished by their thin, sleek body. They are uniform grey above and white below, and their faces have a distinguishing feature: their lower right jaw is white, yet their lower left jaw is black. The reason for this isn't clear – it's just one of the many mysteries of the sea. They feed on krill, squid and small fish such as herring, which they sieve from the water with baleen plates. They open their mouths to take in great gulps of water filled with fish, then close their mouths to filter the water back out through their baleen. In this way fin whales can consumer over 2 tonnes of food a day.

Fin whales are found throughout the ocean, favouring cooler waters, but these majestic giants are by no means as common as they were. Initially they didn't suffer from whaling in the same way that other species did because their incredible swimming speed kept them out of reach. However, the invention of steam-powered ships and explosive harpoons meant that even the ability to sustain speeds of more than 30 kilometres an hour wasn't enough to keep the fin whales safe. With other species becoming rarer, fin whales were a prime target in the twentieth century, and they were killed in their hundreds of thousands. However, it became clear that this hunting was unsustainable and, given the vast distances that whales travel, international action was needed.

The International Whaling Commission was founded to regulate whale hunting, and in 1986 it introduced a global moratorium on commercial whaling. This landmark agreement has reversed the fortunes of fin whales and their relatives. It hasn't ended whaling completely, and permissions for subsistence whaling have allowed the hunting of a very small number of fin whales by Indigenous communities in Greenland. It also hasn't quite put an end to commercial whaling, and Iceland has permitted catches of over 1,500 fin whales since the moratorium was put in place. However, these numbers haven't dented the fin whale's recovery, and there are signs that Iceland's commercial hunts may have come to an end. Public opinion and a lack of demand for whale meat, even to export to Japan, prompted the Icelandic government not to issue permits to catch fin whales in 2019, 2020 or 2021.

The run-up to the 1986 moratorium had already seen a reduction in whaling, but whale populations take time to bounce back. This is to be expected for a long-lived species – the fin whale's standard lifespan is eighty to ninety years, so some of the whales killed by whalers might still have been alive today. One whale killed in Antarctica was found to be 111. Their slow rates of reproduction mean it has taken time to replace the whales that were lost, with each female producing just one calf a year. Still, the population has roughly doubled since the 1970s, now reaching an estimated 100,000.

There are many reasons to celebrate the fin whale's change in fortune. For example, whales play an important role in the ocean food web and carbon storage, as we saw earlier. It is also a perfect example of a synergy between ensuring a thriving population and improving animal welfare. Although killing methods have been improving under the oversight of the International Whaling Commission, they still cause serious concern. Being chased at high speeds is likely to cause the whale stress, just as it would for an animal being chased on land, and it is impossible to ensure that a moving whale has the clean death you would expect for livestock. Whales are hit with harpoons, which then explode inside them, sometimes killing them instantly, but often not. When the whale doesn't die from the first harpoon explosion, a 'secondary killing method' is needed, which can be another harpoon or a rifle. Data reported from Greenland in 2016, for example, showed that fin whales took an average of ten minutes to die after being hit with a harpoon cannon. Not only has the moratorium moved the fin whale further from extinction, it has also vastly reduced this suffering.

Of course the fin whale's problems are far from over. Major threats now include vessel strike, and marine debris such as glass, plastic and abandoned fishing gear. However, momentum is building to change the ways of life that have caused this problem. More people are questioning our use of plastic consumer goods shipped from the other side of the world. The ongoing threats to the fin whale give an insight into what we can achieve by reducing our consumption and changing our habits. There's a long way to go, but there's also evidence that a transformation has begun.

We can see this in our own lives. Plastic bags were once given out liberally for every supermarket shop, something that seems unimaginable now; in India they have been banned altogether. I used to use a facewash with microplastics in, completely oblivious to the fact that these could pass through filtration systems to end up in lakes, rivers and the ocean. The term 'microplastics' was only coined in 2004, yet over the following years awareness of the problem soared. Marine life was the focus of early work, and alarming discoveries included evidence that zooplankton grow more slowly if they ingest microplastics and produce fewer larvae.

There are still unanswered questions about the impact of microplastics on humans and wildlife, but the growing evidence of harm means that action has already been taken. In 2015, the USA banned plastic microbeads in cosmetics and personal care products, and the UK followed suit in 2018 with a ban on microbeads in rinse-off cosmetics. Other countries banning microbeads range from Sweden to New Zealand. The problem is by no means solved, with microplastics being shed by everything from car tyres to synthetic clothing. However, it's important to acknowledge and celebrate the progress that has been made on the intractable problem of plastic pollution. According to a 2018 UN report, over sixty countries have introduced regulations to limit or ban the use of disposable plastics. One day we will look back at today's world and be incredulous about the amount of plastic we used and discarded so lightly.

Despite the scale of plastic waste, the ocean arguably faces an even greater global problem: the climate crisis. The fin whale may have had a change of fortune, but scientists have warned that our current attempts to protect them may fail unless we tackle our carbon emissions. Warming will have a devastating effect on many species, and the casualties are likely to include the krill that fin whales feed on. It's another reminder that, however hard we work to protect species and habitats, high numbers of extinctions are inevitable if we experience a catastrophic event such as nuclear war or extreme climate change. Climate change predictions suggest that even the most likely scenario would bring about the demise of a wide range of species, and we should also be mindful that there's the potential for much worse. Scientists at the University of Cambridge's Centre for Existential Risk have given an alarming warning: if we continue with a medium–high emissions pathway, they estimate that there is a 10 per cent chance that eventual warming will be 6 °C, and a 3 per cent chance of 10 °C warming.

Such calamities need not happen, and everyone who plays a role in reducing them will have helped to prevent extinctions and reduce suffering; every reduction we make in the peak temperature has huge benefits for humans and the rest of nature. With political will lagging behind public concern, it is easy to feel that we are fighting a losing battle. But again there are signs of hope. Polls in 2020 revealed that two thirds of British people believed that, in the long term, climate change is as serious a crisis as Covid-19. In other countries the number is higher: 87 per cent of Chinese believe this, as do 84 per cent of Mexicans and 81 per cent of Indians. People are demanding action, which has sometimes gone alongside demands to rethink democracy.

In Chapter 8, we saw how conservation challenges could be addressed by bringing a diversity of people together. However, the troubles in the Moray Firth were addressed with a small discussion group, whereas a global problem needs global solutions. If these global solutions are going to be fair, we can't exclude any groups of people. Throughout this book we have seen that there are no right answers to questions about the natural world, which opens up opportunities for everyone to contribute creative ideas, rather than a narrow range of experts searching for a solution.

This acknowledgement that we need diverse voices led to the announcement of a global citizens' assembly to be held in the run-up to the 2021 UN climate talks in Glasgow (COP26). One hundred participants from around the world were chosen by lottery to deliberate the question 'How can humanity address the climate and ecological crisis in a fair and effective way?'

This idea of citizens' assemblies, inspired by the original means of Athenian democracy, has appeared in different forms this century. In 2015, Ireland's Convention on the Constitution selected sixty-six citizens and thirty-five politicians to spend a year in consultation, tackling issues that had long been divisive. The assembly's recommendations paved the way for policy change, and led to the Irish public voting in favour of gay marriage in a national referendum. Abortion was likewise legalised following a subsequent citizens' assembly. Other recommendations never materialised as policy. In the UK, a citizen's assembly discussed Brexit in 2017, and came to a much more nuanced view than public debates, notably on immigration.

The drive to hold citizens' assemblies about climate change comes partly from Extinction Rebellion (XR), a global environmental movement that started in the UK in 2018. In response to their demands, the Climate Assembly UK brought together 108 people who were representative of the UK population, both demographically and in levels of concern about climate change. The assembly convened for six weekends in early 2020 to explore how the UK should reach its target of net zero greenhouse gas emissions by 2050. They heard a wide range of views and evidence before engaging in in-depth discussion about the best way forward. Their recommendations were varied and bold, ranging from cutting meat consumption to banning the sale of petrol and diesel cars by 2030–2035. They supported some standard solutions and some creative ones, such as a suggestion that a tax on air travel should rise as people fly more – your fifth flight of the year would be taxed much more than your first. Although the government isn't legally bound to follow the Assembly's recommendations, some have been implemented.

The process isn't perfect, and can't be free from political influence. The organisers choose the experts who speak and the topics they cover, introducing a potential source of bias. There will inevitably be a lack of diversity among experts, reflecting biases in society. For example, women made up less than a quarter of authors on the 2013 Intergovernmental Panel on Climate Change report. And a recent study by Media Matters for America found that people of colour accounted for less than 10 per cent of people interviewed or featured in media coverage on climate change. Likewise, the Climate Assembly UK discussion was limited to the target that had been set by politicians: net zero carbon emissions by 2050. However, the very fact that it took place and produced thoughtful recommendations gives hope for a future where climate justice is taken seriously. Climate change will hit the poorest hardest and soonest, and finally it seems as if their voices might be heard.

As we saw in Chapter 6, the fate of the natural world is intimately tied to the way we produce our food. It's clear that the current food system has an unacceptable impact on wildlife, and also leaves us vulnerable. A narrow focus on conservation risks exacerbating the problems rather than solving them. Protecting biodiversity hotspots from conversion to agriculture, for example, can lead to food price spikes, with the poorest people hit the hardest. However, integrating policies for wildlife conservation with support for a transition to sustainable agriculture, diets and lifestyles can simultaneously protect nature and enhance human well being. One example comes from the Red Siskin Initiative in Venezuela, which shows how actions to protect a charismatic bird species can have wide benefits for wildlife and people.

Red siskins used to be widespread in northern Venezuela, but their numbers have plummeted in the last 100 years so that they are now confined to a few isolated populations. These beautiful birds have deep orange-red bodies which contrast with their black head and wings, a striking coloration that has made them popular with breeders, who cross them with domestic canaries to create desired 'red factor' birds. Although trade of the red siskin is prohibited under the Convention on International Trade in Endangered Species of Wild Fauna and Flora (CITES), the practice still continues. Red siskins are caught from the wild to be trafficked in South America and to Spain, with consequences both for the population and for animal welfare.

The Red Siskin Initiative is another example of where animal welfare and conservation goals intersect – they rescue birds from illegal trade and rehabilitate them with the aim of returning them to the wild. If the red siskin is going to survive in the wild, though, it needs somewhere to go. Alongside the threat from the caged bird trade, the species has also suffered from the loss of its forest habitat, particularly due to logging, agriculture and urbanisation. The Red Siskin Initiative therefore works with local people to preserve the tropical dry forests of northern Venezuela. Rather than being excluded from the forest, local people are taught agroforestry techniques to grow coffee, bananas and other fruits within the forest. By supporting this sustainable use of the forest, we move on from an ideology that separates humans from nature or aspires to return the forest to a past baseline. An incredible diversity of birds has been seen in the agroforestry plantations, and farmers have the opportunity to grow quality products they can sell for a premium. This is a far cry from both the unsustainable agriculture that leads to forest clearance and the fortress conservation policies that remove people from the forest that supports them.

This programme is just one example of the many situations where farming transformations are happening due to increased knowledge rather than increased inputs. Community-based natural farming is taking hold around India, for example, with government policies contributing to its rapid uptake. This method of farming promotes the use of locally-sourced inputs such as cow dung, and farmers grow a greater diversity of crops. State governments are providing training on a huge scale, but this is about more than passing on information – it is about empowering people. Support groups for women have been particularly important, as has their increased access to microfinance. Whereas volatile food prices and the high cost of artificial inputs had pushed farmers into debt they couldn't repay, microfinance is providing access to cheaper, more sustainable alternatives. The transformation in India inspires hope that agriculture around the world can shift to have people and sustainability at its heart, not power and profit.

There are many ways we can be part of this revolution, whether this is growing vegetables in a community garden, reducing our consumption of animal products, or giving microfinance loans through organisations such as Kiva.

There are also many ways we can directly support nature conservation, through volunteering, donations or activism. We all have the power to create a world with more nature

– you can start your own miniature rewilding project with nothing more than a window box. The path forwards won't be easy, but it is filled with opportunities. Now that we have shaken off the idea of an unobtainable 'pure' nature, we can embrace the possibilities that come with celebrating nature in its many forms. We're not doomed to simply mourn a paradise lost, we're free to create a new form of paradise. If we shift our values to cherish life and not the consumerism that destroys it, and approach our choices with wisdom and compassion, then a brighter future is within our grasp.

SOURCES AND ENDNOTES

CHAPTER 1

Nikita Zimov has a provided a huge amount of information on the Pleistocene Park website, including a video documentary: https://pleistocenepark.ru. He has co-authored multiple papers about Pleistocene Park, including Macias-Fauria et al. (2020) 'Pleistocene Arctic megafaunal ecological engineering as a natural climate solution?' *Philosophical Transactions of the Royal Society B,* 375, 20190122.

Sofia Castelló y Tickell has explored human structures at sea in her 2019 paper 'Sunken worlds: the past and future of human-made reefs in marine conservation' *BioScience,* 69, 725–735. This includes information about the ancient rock terraces of British Columbia.

Helen Pilcher's book *Life Changing* (2021) gives an excellent insight into ways that humans are altering life on Earth.

For an analysis of the history of Heck cattle and their use in rewilding projects, see Lormier et al. (2016) 'From "Nazi cows" to cosmopolitan "ecological engineers": specifying rewilding through a history of Heck cattle' *Annals of the American Association of Geographers,* 106, 631–652.

Theodore Roosevelt was a prolific writer, and *Hunting for the Grizzly and Other Stories* (1900) is just one of the books that can give an insight into views on wilderness at the turn of the twentieth century. A more contemporary analysis can be found in 'For wilderness or wildness? decolonising rewilding' by Kim Ward, a chapter in Pettorelli et al. (ed) *Rewilding* (2019).

CHAPTER 2

Serian Sumner and colleagues have published extensively on the benefits of wasps and our learnings from the public submission of data. See in particular: Sumner et al. (2019) 'Mapping species distributions in 2 weeks using citizen science' *Insect Conservation and Diversity*, 12, 382–388; and Brock et al. (2021) 'Ecosystem services provided by aculeate wasps' *Biological Reviews*, 96, 1645–1675.

If you want to explore how honeybee numbers are changing globally or in a specific country, the UN's Food and Agricultural Organisation has a comprehensive database which allows you to draw your own graphs: www.fao.org/faostat/en/#data/QCL.

Free-living honeybees are rare in Europe, but small populations exist that are genetically distinct from domesticated honeybees. For more information about these wild honeybees, see Browne et al. (2021) 'Investigation of free-living honey bee colonies in Ireland'. *Journal of Apicultural Research*, 60, 229–240. For thoughts on their conservation, see Requier et al. (2019) 'The conservation of native honey bees is crucial' *Trends in Ecology and Evolution*, 34, 789–798.

The Intergovernmental Platform on Biodiversity and Ecosystem Services (IPBES) has a wealth of resources. Of particular interest is their *Assessment Report on Pollinators, Pollination and Food Production*. There are also many papers coming out with analysis of insect declines, including Halsch et al. (2021) 'Insects and recent climate change' *Proceedings of the National Academy of Sciences of the United States of America*, 18(2), e2002543117.

For more information about the impacts of parasites, see Sato et al. (2019) 'Host manipulation by parasites as a cryptic driver of energy flow through food webs' *Current Opinion in Insect Science*, 33, 69–76, or Hughes et al. *(2013) Host Manipulation by Parasites.*

A summary of European conservation biases can be found in Mammides (2019) 'European Union's conservation efforts are taxonomically biased' *Biodiversity and Conservation*, 28, 1291–1296.

For a review of how an understanding of host–microbe interactions can benefit conservation, see Carthey (2020) 'Conserving the holobiont' *Functional Biology,* 4, 764–776.

For more information about *Mycobacterium vaccae* and the health impact of microorganisms, a thorough review can be found in Robinson (2020) 'Rekindling old friendships in new landscapes: the environment–microbiome–health axis in the realms of landscape research' *People and Nature,* 2, 339–349.

The warning I referred to about the microscopic majority can be found in Cavicchioli et al. (2019) 'Scientists' warning to humanity: microorganisms and climate change' *Nature Reviews Microbiology,* 17, 569–586.

Diana Wall, mentioned in relation to *Scottnema lindsayae,* has produced fascinating research on the hidden wildlife of Antarctica. A particularly interesting paper she contributed to is Gutt et al. (2021) 'Antarctic ecosystems in transition – life between stresses and opportunities' *Biological Reviews, 96, 798–821.*

CHAPTER 3

Sara Busilacchi's work in Papua New Guinea is worth a read, including Butler et al. (2019) 'How resilient is the Torres Strait Treaty (Australia and Papua New Guinea) to global change? A fisheries governance perspective' *Environmental Science & Policy,* 91, 17–26.

Chris Thomas's very readable book *Inheritors of the Earth* (2017) shares positive thoughts on the ways that humans have changed nature, particularly in relation to introduced species. Likewise, Ken Thomas's book *Where do Camels Belong?* (2014) is a fascinating exploration of why few introduced species become established and even fewer of them become a problem.

For a review on the impact of non-native species and ways we can deal with this, see Hanley et al. (2019) 'The economic benefits of invasive species management' *People and Nature,* 1, 124–137.

For a thorough analysis of ring-necked parakeets and other parrots in Europe, see White et al. (2019) 'Assessing the ecological and societal impacts of alien parrots in Europe using a transparent and inclusive evidence-mapping scheme' *NeoBiota,* 48, 45–69.

CHAPTER 4

The paper I mention in relation to the Guam rail louse is Rózsa et al. (2015) 'Co-extinct and critically co-endangered species of parasitic lice, and conservation-induced extinction: should lice be reintroduced to their hosts?' *Oryx,* 49, 107–110.

For further information about brown treesnake control, see McEldery et al. (2021) 'Distilling professional opinion to gauge vulnerability of Guam avifauna to brown treesnake predation' *Frontiers in Conservation Science,* https://doi.org/10.3389/fcosc.2021.683964

For an in-depth exploration of the philosophy of environmental ethics and different ways of valuing nature, Newman et al.'s (2017) *Defending Biodiversity* makes excellent reading. I have provided a more succinct exploration of values on my website (https://rebeccanesbit.com/what-is-the-point-of-conservation/).

More information the crop wild relatives can be found in Volk and Byrne (ed.) *Crop Wild Relatives and Their Use in Plant Breeding* (2020). Volk et al.'s chapter 'Introduction to crop relatives' gives an insight into Vavilov's life.

An insight into the possibility of transgenic chestnuts can be found in Popkin (2018) 'Can a transgenic chestnut restore a forest icon?' *Science,* 361, 830–831. And, for a broader perspective, see Isabel et al. (2020) 'Forest genomics: advancing climate adaptation, forest health, productivity, and conservation' *Evolutionary Applications,* 13, 3–10.

The possibility of changing scarlet honeycreeper genetics is explored in Samuel et al. (2020) 'Facilitated adaptation for conservation – can gene editing save Hawaii's endangered birds from climate driven avian malaria?' *Biological Conservation,* 241, 108390.

CHAPTER 5

Greg Howald and collaborators have produced extensive research and commentary related to control of non-native animals, including: de Wit et al. (2020) 'Invasive vertebrate eradications on islands as a tool for implementing global Sustainable Development Goals' *Environmental Conservation,* 47, 139–148; and Dubois et al. (2017).

'International consensus principles for ethical wildlife control' *Conservation Practice and Policy,* 31, 753–760.

Ongoing updates about the Floreana Restoration Project can be found on the Island Conservation and Galapagos Conservation Trust websites (www.islandconservation.org, www.galapagos conservation.org.uk).

My telling of the vegan fox thought experiment was adapted from Horta (2010) 'Disvalue in nature and intervention' *Pensata Animal,* 34.

Jeremy Lormier's book *Wildlife in the Anthropocene: Conservation after Nature* (2016) offers fascinating insights into changing definitions of nature and the politics of conservation, with particular insights into Oostvaardersplassen. For an alternative perspective, see Theunissen (2019) 'The Oostvaardersplassen fiasco' *Isis,* 110, https://doi.org/10.1086/703338.

Isabella Tree shared the story of how she transformed the Knepp Estate in her 2018 book *Wildling: The Return of Nature to a British Farm.*

The report of the ex situ options for cetacean conservation workshop, in Nuremberg, Germany, 2018, is publicly available from the IUCN: https://portals.iucn.org/library/sites/library/files/documents/SSC-OP-066-En.pdf.

CHAPTER 6

For an analysis of the impact of Half-Earth, see Schleicher et al. (2019) 'Protecting half of the planet could directly affect over one billion people' *Nature Sustainability,* 2, 1094–1096.

The history and findings of Park Grass can be found in Silvertown (2006) 'The Park Grass Experiment 1856–2006: its contribution to ecology' *Journal of Ecology,* 94, 801–814.

Anyone interested in the sharing vs sparing debate will find a wealth of information in the work of Andrew Balmford, Tom Finch and collaborators. One recent paper worth reading is Balmford (2021) 'Concentrating vs. spreading our footprint: how to meet humanity's needs at least cost to nature' *Journal of Zoology,* 315, 79–109.

The Wildlife Trusts and the University of Essex have produced a through report about the impact of nature on human health: Bragg et al. (2018) 'Wellbeing benefits from natural environments rich in wildlife', www.wildlifetrusts.org/sites/default/files/2018-05/r1_literature_review_wellbeing_benefits_of_wild_places_lres_0.pdf.

More information about sheep grazing in Norway can be found in Austrheim et al. (2016) 'Synergies and trade-offs between ecosystem services in an alpine ecosystem grazed by sheep – an experimental approach' *Basic and Applied Ecology,* 17, 596–608.

A detailed insight into the situation faced by the Trio in Suriname can be found in Hill et al. (ed.) (2017) *Understanding Conflicts about Wildlife: A Biosocial Approach.*

CHAPTER 7

There is a large and growing literature on protected areas and social justice. One interesting paper to which David Obura contributed is Bennett et al. (2021) 'Advancing social equity in and through marine conservation' *Frontiers in Marine Science,* doi.org/10.3389/fmars.2021.711538.

Two papers with more information about the Phoenix Islands Protected Area are: Rotjan et al. (2014) 'Establishment, management, and maintenance of the Phoenix Islands Protected Area' Advances in *Marine Biology,* 69, 289–324; and Mallin et al. (2019) 'In oceans we trust: conservation, philanthropy, and the political economy of the Phoenix Islands Protected Area' *Marine Policy,* 107, 103421.

For anyone interested in sharks, a thorough introduction can be found in Abel and Grubbs (2020) *Shark Biology and Conservation: Essentials for Educators, Students, and Enthusiasts.*

For discussion about Gorongosa as a case study for effective protected areas, see Pringle (2017) 'Upgrading protected areas to conserve wild biodiversity' *Nature,* 546, 91–99.

Gorongosa community engagement, with a focus on the impact of elephants from the park raiding nearby croplands, is explored in Branco et al. (2019) 'An experimental test of community-based strategies for mitigating human–wildlife conflict around protected areas' *Conservation Letters*, e12679.

For insights into the situation in Kaziranga, see: Barbora (2017) 'Riding the rhino: conservation, conflicts, and militarisation of Kaziranga National Park in Assam' *Antipode,* 49, 1145–1163; and Smadja (2018) 'A chronicle of law implementation in environmental conflicts: the case of Kaziranga National Park in Assam (North-East India)' *South Asia Multidisciplinary Academic Journal*, 17, doi.org/10.4000/samaj.4422.

CHAPTER 8

James Butler and Juliette Young continue to make contributions to the field of conservation conflicts, including Butler et al. (2021) 'Decision-making for rewilding: an adaptive governance framework for social-ecological complexity' *Frontiers in Conservation Science*, 2, article 681545. They have also published extensively about the Moray Firth, including: Butler et al. (2011) 'Perceptions and costs of seal impacts on Atlantic salmon fisheries in the Moray Firth, Scotland: Implications for the adaptive co-management of seal–fishery conflict' *Marine Policy,* 35, 317–323.

The scientific literature about trophy hunting is extensive and varied. Of particular interest may be Mkono (2018) 'Neo-colonialism and greed: Africans' views on trophy hunting in social media' *Journal of Sustainable Tourism,* 27, 689–704.

For an example of environmental injustice in cities, see Schell (2020) 'The ecological and evolutionary consequences of systemic racism in urban environments' *Science,* 369, doi.org/10.1126/science.aay4497.

For more information about vultures, see Safford et al. (2019) 'Vulture conservation: the case for urgent action' *Bird Conservation International*, 29, 1–9.

Angela Cassidy has provided thoughtful analysis of the debate over badgers in the UK, including in her 2019 book *Vermin, Victims and Disease: British Debates over Bovine Tuberculosis and Badgers.*

EPILOGUE

There is a broad literature on responding to the crises we face, and how we can feed the global population while protecting nature. Some examples are: Balmford et al. (2020) 'Analogies and lessons from COVID-19 for tackling the extinction and climate crises' *Current Biology*, 30, R963–R983; Leclère et al. (2020) 'Bending the curve of terrestrial biodiversity needs an integrated strategy' *Nature*, 585, 551–556; and Pascual et al. (2021) 'Biodiversity and the challenge of pluralism' *Nature Sustainability*, 4, 567–572.

Information about the woodrose can be found in Holzapfel (2016) 'Successful translocation of the threatened New Zealand root-holoparasite *Dactylanthus taylorii (Mystropetalaceae)*' *Plant Ecology*, 217, 127–138.

For a warning about whale recovery, see Tulloch et al. (2019) 'Future recovery of baleen whales is imperiled by climate change' *Global Change Biology*, 25, 1263–1281. For an exploration of the ethics of whaling, see Garner (2011) 'Animal welfare, ethics and the work of the International Whaling Commission' *Journal of Global Ethics*, 7, 279–290.

For more about the 2021 global citizens' assembly, visit https://globalassembly.org; and, for Climate Assembly UK, see www.climateassembly.uk. A readable introduction to the citizens' assembly is David Van Reybrouck's (2016) *Against Elections: The Case for Democracy*.

PHOTO CREDITS

Introduction **WHY PROTECT NATURE?**
Male silverback gorilla
Parc National des Volcans, Rwanda
Sajjad Hussain/AFP via Getty Images

Chapter 1 **THE MYTH OF WILD NATURE**
Adult Zubr or European bison
Chemalsky District, Altai Republic, Russia
Gekko Studios/Alamy Stock Photo

Chapter 2 **THE BEAUTIES AND THE BEASTS**
Yellowjacket wasp
Antagain/Getty Images

Chapter 3 **NEW ARRIVALS**
Mozambique Tilapia
Ammit/Alamy Stock Photo

Chapter 4 **NOAH'S ARK**
Guam rail bird
Guam
Guam Endangered Species Recovery Program

Chapter 5 **ANIMAL WELFARE**

Galápagos giant tortoise and brown rat
Floreana, Galápagos
Krystyna Szulecka/Alamy Stock Photo

Chapter 6 **A HUMAN LANDSCAPE**

Yellowhammer
Norfolk, England
Ernie Janes/Alamy Stock Photo

Chapter 7 **FORTRESS CONSERVATION**

Grey reef shark and whitetip reef shark
Papua New Guinea
Helmut Corneli/Alamy Stock Photo

Chapter 8 **WINNERS AND LOSERS**

Grey seal with a half-eaten salmon in its mouth
Scotland
Alamy Stock Photo

Epilogue **SHAPING OUR FUTURE**

The tail of a fin whale
Gulf of California
Biosphoto/Alamy Stock Photo

THANKS

I am very grateful to the many scientists and conservationists who have given their time and shared their knowledge freely. In addition to the people mentioned in the text, they include Jonathan Newman, Brandon Keim, John Ewen, Mathilde Poyet, Skylar Hopkins, Tom Finch, Alex Caveen, Steve Widdicombe, Jen Jones, Paul Somerfield, Carol Turley, Nathalie Pettorelli, Duncan Vaughan, Dean Grubbs, Fergus Kennedy, James Speed, Roberto Zolho, Tomas Chaigneau, Jake Robinson, Adam Hart and Rose Moorhouse-Gann. The support I received from the British Ecological Society was fundamental to bringing my idea for a book to fruition, with particular thanks to Kate Harrison. I received insightful feedback from Mark Ellingham at Profile Books and Ken Thompson, who have both made this a better book. Thanks also to Nikky Twyman for proofreading, Henry Iles for design and Bill Johncocks for the index.

I am ever grateful to the family and friends who have supported me at every stage, from nurturing my love of nature to editing the final book. There are far too many people to thank by name, but I should particularly mention my mother, who always provides astute feedback on early drafts, and my husband Phil, who has provided endless encouragement.

INDEX

The index covers the main text but not the Sources. *Italic* page numbers indicate the start-of-chapter illustrations. Many species are listed at their inverted common names (e.g. both yellow-legged and herring gulls are at 'g') but others are gathered as families (cephalopods, cetaceans, owls...).

243